Einstein's Miraculous Year

Einstein's bookplate, drawn by Erich Büttner. (Courtesy of Hebrew
University of Jerusalem)

Einstein's Miraculous Year

✦

*Five Papers That Changed
the Face of Physics*

EDITED AND INTRODUCED BY
John Stachel

WITH THE ASSISTANCE OF
Trevor Lipscombe, Alice Calaprice,
and Sam Elworthy

AND

WITH A FOREWORD BY
Roger Penrose

PRINCETON UNIVERSITY PRESS

Library of Congress Cataloging-in-Publication Data
Einstein's miraculous year : five papers that changed the
face of physics / edited and introduced by John Stachel ;
with the assistance of Trevor Lipscombe,
Alice Calaprice, Sam Elworthy ;
with a foreword by Roger Penrose.
p.   cm.
ISBN 0-691-05938-1 (cloth : alk. paper)
1. Physics—History—20th century. 2. Einstein, Albert,
1879–1955. I. Stachel, John J., 1928–   .
QC7.E52   1998
530.1—dc21   97-48441

CONTENTS

FOREWORD

In the twentieth century, we have been greatly privileged to witness *two* major revolutions in our physical picture of the world. The first of these upturned our conceptions of space and time, combining the two into what we now call *space-time*, a space-time which is found to be subtly curved in a way that gives rise to that long-familiar, omnipresent but mysterious, phenomenon of gravity. The second of these revolutions completely changed the way in which we understand the nature of matter and radiation, giving us a picture of reality in which particles behave like waves and waves like particles, where our normal physical descriptions become subject to essential uncertainties, and where individual objects can manifest themselves in several places at the same time. We have come to use the term "relativity" to encompass the first of these revolutions and "quantum theory" to encompass the second. Both have now been observationally confirmed to a precision unprecedented in scientific history.

I think that it is fair to say that there are only three previous revolutions in our understanding of the physical world that can bear genuine comparison with either. For the first of these three, we must turn back to ancient Greek times, where the notion of Euclidean geometry was introduced and some conception was obtained of rigid bodies and static configurations. Moreover, there was a beginning of an appreciation of the crucial role of *mathematical reasoning* in our insights into Nature. For the second of the three, we must leap to the seventeenth century, when Galileo and Newton told us how the motions of ponderable bodies can be understood in terms of forces between their constituent particles and the accelerations that these forces engender. The nineteenth century gave us the third revolution, when Faraday and Maxwell showed us that particles were not enough, and we must consider, also, that there are continuous fields pervading space, with a reality as great as that of the particles themselves. These fields were combined into a single all-pervasive entity, referred to as the *electromagnetic field*, and the behavior of light could be beautifully explained in terms of its self-propagating oscillations.

Turning now to our present century, it is particularly remarkable that a single physicist—Albert Einstein—had such extraordinarily deep perceptions of the workings of Nature that he laid foundation stones of *both* of these twentieth-century revolutions in the *single year* of 1905. Not only that, but in this same year Einstein also provided fundamental new insights into two other areas, with his doctoral dissertation on the determination of molecular dimensions and with his analysis of the nature of Brownian motion. This latter analysis alone would have earned Einstein a place in history. Indeed, his work on Brownian motion (together with the independent and parallel work of Smoluchowski) laid the foundations of an important piece of statistical understanding which has had enormous implications in numerous other fields.

This volume brings together the five papers that Einstein published in that extraordinary year. To begin with, there is the one just referred to on molecular dimensions (paper 1), followed by the one on Brownian motion (paper 2). Then come two on the special theory of relativity: the first initiates the "relativity" revolution, now so familiar to physicists (and also perceived by the public at large), in which the notion of absolute time is abolished (paper 3); the second is a short note deriving Einstein's famous "$E = mc^2$" (paper 4). Finally, the (only) paper that Einstein himself actually referred to as "revolutionary" is presented, which argues that we must, in some sense, return to the (Newtonian) idea that light consists of *particles* after all—just when we had become used to the idea that light consists solely of electromagnetic waves (paper 5). From this apparent paradox, an important ingredient of quantum mechanics was born. Together with these five classic Einstein papers, John Stachel has provided fascinating and highly illuminating introductions that set Einstein's achievements in their appropriate historical settings.

I have referred above to the twentieth century's two extraordinary revolutions in physical understanding. But it should be made clear that, fundamental as they were, Einstein's papers of 1905 did not quite provide the initial shots of those revolutions; nor did these particular papers set out the final nature of their new regimes.

The revolution in our picture of space and time that Einstein's two 1905 relativity papers provided concerned only what we now call the *special* theory. The full formulation of the *general* theory of relativity, in which gravitation is interpreted in terms of curved space-time geometry, was not achieved until ten years later. And even for special relativity, the wonderful insights presented by Einstein in 1905 provided a theory that was not totally original with him, this theory having been grounded in earlier ideas (notably those of Lorentz and Poincaré). Moreover, Einstein's viewpoint in 1905 still lacked one important further insight—that of *space-time*—introduced by Hermann Minkowski three years later. Minkowski's notion of a four-dimensional space-time was soon adopted by Einstein, and it became one of the crucial steppingstones to what was later to become Einstein's crowning achievement: his *general* theory of relativity. With regard to quantum mechanics, the initial shots of this revolution had been Max Planck's extraordinary papers of 1900, in which the famous relation $E = hv$ was introduced, asserting that energy of radiation is produced in discrete little bundles, in direct proportion to the radiation's frequency. But Planck's ideas were hard to make sense of in terms of the ordinary physics of the day, and only Einstein seems to have realized (after some while) that these tentative proposals had a fundamental significance. Quantum theory itself took many years to find its appropriate formulation— and this time the unifying ideas came not from Einstein, but from a number of other physicists, most notably Bohr, Heisenberg, Schrödinger, Dirac, and Feynman.

There are some remarkable aspects to Einstein's relation to quantum physics, which border almost on the paradoxical. Earliest and perhaps most striking of these seeming paradoxes is the fact that Einstein's initial revolutionary papers on quantum phenomena (paper 5) and on relativity (paper 3) appear to start from mutually contradictory standpoints with regard to the status of Maxwell's electromagnetic theory as an explanation of light. In paper 5, Einstein explicitly rejects the view that Maxwell's equations suffice to explain the actions of light (as waves in the electromagnetic field) and he puts forward a model in which light behaves, instead, like little particles. Yet, in (the later) paper 3, he develops the special theory of relativity from the starting point that Maxwell's theory indeed *does* represent fundamental truth, and the relativity theory that Einstein constructs is specifically de-

signed so that it leaves Maxwell's equations intact. Even at
the beginning of paper 5 itself, where Einstein puts forward
a "particle" viewpoint of light in conflict with Maxwell's theory, he comments on the latter (wave) theory of light that
it "will probably never be replaced by another theory." This
seeming conflict is all the more striking when one considers
that Einstein's incredible strength as a physicist came from
his direct physical insights into the workings of Nature. One
could well imagine some lesser figure "trying out" one model
and then another (as is common practice with physicists of
today), where the contradiction between the two proposed
viewpoints would cause no real concern, since neither carries any particular conviction. But with Einstein, things were
quite different. He appears to have had very clear and profound ideas as to what Nature was "really like" at levels not
readily perceivable by other physicists. Indeed, his ability to
perceive Nature's reality was one of his particular strengths.
To me, it is virtually inconceivable that he would have put
forward two papers in the same year which depended upon
hypothetical views of Nature that he felt were in contradiction with each other. Instead, he must have felt (correctly,
as it turned out) that "deep down" there was no real contradiction between the accuracy—indeed "truth"—of Maxwell's
wave theory and the alternative "quantum" particle view that
he put forward in paper 5.

One is reminded of Isaac Newton's struggles with basi-
cally the same problem—some 300 years earlier—in which
he proposed a curious hybrid of a wave and particle viewpoint in order to explain conflicting aspects of the behavior
of light. In Newton's case, it is possible to understand his
dogged adherence to a particle-type picture if one takes the
(reasonable) view that Newton wished to preserve a relativ-

ity principle. But this argument holds only if the relevant relativity principle is that of Galileo (and Newton). In Einstein's case, such an argument will not do, for the reason that he explicitly put forward a *different* relativity principle from the Galilean one, in which Maxwell's wave theory could survive intact. Thus, it is necessary to look more deeply to find the profound reasons for Einstein's extraordinary conviction that although Maxwell's wave picture of light was, in some sense, "true"—having been well established in 1905— it nevertheless needed to be altered to something different which, in certain respects, harked back Newton's hybrid "wave-particle" picture of three centuries earlier.

It would seem that one of the important influences that guided Einstein was his awareness of the conflict between the discrete nature of the particles constituting ponderable bodies and the continuous nature of Maxwell's fields. It is particularly manifest in Einstein's 1905 papers that this conflict was very much in his mind. In papers 1 and 2, he was directly concerned with demonstrating the nature of the molecules and other small particles which constitute a fluid, so the "atomic" nature of matter was indeed at the forefront. In these papers, he showed himself to be a master of the physical/statistical techniques required. In paper 5, he put this extraordinary expertise to use by treating electromagnetic fields in the same way, thereby explaining effects that cannot be obtained with the Maxwellian view of light alone. Indeed, it was made clear by Einstein that the problem with the classical approach was that a picture in which continuous fields and discrete particles *coexist*, each interacting with the other, does not really make physical sense. Thus, he initiated an important step toward the present-day quantum-theoretic viewpoint that particles must indeed take on attributes of waves, and fields must take on attributes of particles. Looked at appropriately in the quantum picture, particles and waves actually turn out to be the same thing.

The question is often raised of another seeming paradox: Why, when Einstein started from a vantage point so much in the lead of his contemporaries with regard to understanding quantum phenomena, was he nevertheless left behind by them in the subsequent development of quantum theory? Indeed, Einstein never even accepted the quantum theory, as that theory finally emerged in the 1920s. Many would hold that Einstein was hampered by his "outdated" *realist* standpoint, whereas Niels Bohr, in particular, was able to move forward simply by denying the very existence of such a thing as "physical reality" at the quantum level of molecules, atoms, and elementary particles. Yet, it is clear that the fundamental advances that Einstein was able to achieve in 1905 depended crucially on his robust adherence to a belief in the *actual* reality of physical entities at the molecular and submolecular levels. This much is particularly evident in the five papers presented here.

Can it really be true that Einstein, in any significant sense, was as profoundly "wrong" as the followers of Bohr might maintain? I do not believe so. I would, myself, side strongly with Einstein in his belief in a submicroscopic reality, and with his conviction that present-day quantum mechanics is fundamentally incomplete. I am also of the opinion that there are crucial insights to be found as to the nature of this reality that will ultimately come to light from a profound analysis of a seeming conflict between the underlying principles of quantum theory and those of Einstein's own general relativity. It seems to me that only when such insights are at hand and put appropriately to use will the fundamental tension between the laws governing the micro-world of quantum theory and the macro-world of general relativity be resolved. How is this resolution to be achieved? Only time and, I believe, a *new* revolution will tell—in perhaps some other *Miraculous Year*!

Roger Penrose
December 1997

PUBLISHER'S PREFACE

In 1905, Einstein produced five of his most significant contributions to modern science, all of which first appeared in the prestigious German journal *Annalen der Physik* in that year. More recently, they have reappeared in the original German, with editorial annotations and prefatory essays, in volume 2 of the *Collected Papers of Albert Einstein*, an ongoing series of volumes being prepared by the Einstein Papers Project at Boston University under the sponsorship of Princeton University Press and the Hebrew University of Jerusalem.

Einstein's Miraculous Year draws heavily from this volume (*The Swiss Years: Writings, 1900–1909*), which remains the definitive and authoritative text of all of Einstein's writings of those years; we encourage scholars to consult it when seeking original texts and detailed discussions and annotations of Einstein's work. For the present volume, we have compiled Einstein's five major papers of 1905 and included, in abridged form, the historical essays and notes that deal with his contributions to relativity theory, quantum mechanics, and statistical mechanics and adapted them for presentation in this special edition. We are therefore indebted to the editors of volume 2 for their scholarly contributions: John Stachel, David C. Cassidy, A. J. Kox, Jürgen Renn, and Robert Schulmann.

The English translations that appear here are new. The intention has been to render Einstein's scientific writings accurately into modern English, but to retain the engaging and clear prose style of the originals. We are deeply grateful to Trevor Lipscombe, Alice Calaprice, Sam Elworthy, and John Stachel for preparing them.

Einstein's Miraculous Year

INTRODUCTION

I

To anyone familiar with the history of modern science, the phrase "miraculous year" in the title immediately calls to mind its Latin counterpart "annus mirabilis," long used to describe the year 1666, during which Isaac Newton laid the foundations for much of the physics and mathematics that revolutionized seventeenth-century science. It seems entirely fitting to apply the same phrase to the year 1905, during which Albert Einstein not only brought to fruition parts of that Newtonian legacy, but laid the foundations for the break with it that has revolutionized twentieth-century science.

But the phrase was coined without reference to Newton. In a long poem entitled *Annus Mirabilis: The Year of Wonders, 1666*, John Dryden, the famed Restoration poet, celebrated the victory of the English fleet over the Dutch as well as the city of London's survival of the Great Fire. The term was then used to celebrate Newton's scientific activities during the same year—a year in which he laid the foundations of his version of the calculus, his theory of colors, and his theory of gravitation.[1] Here is Newton's own (much later) summary of his accomplishments during this period:

> In the beginning of the year 1665 I found the Method of approximating series & the Rule for reducing any dignity

[power] of any Binomial into such a series [i.e., the binomial theorem]. The same year in May I found the method of Tangents . . . , & in November had the direct method of fluxions [i.e., the differential calculus] & the next year in January had the Theory of colours & in May following I had entrance into [th]e inverse method of fluxions [i.e., the integral calculus]. And the same year I began to think of gravity extending to [th]e orb of the Moon & (having found out how to estimate the force with w[hi]ch [a] globe revolving within a sphere presses the surface of a sphere [i.e., the centrifugal force]): from Kepler's rule of the periodical times of the Planets being in sesquialterate proportion of their distances from the centers of their Orbs [i.e., Kepler's third law], I deduced that the forces w[hi]ch keep the Planets in their Orbs must [be] reciprocally as the squares of their distances from the centers about w[hi]ch they revolve: & thereby compared the force requisite to keep the Moon in her Orb with the force of gravity at the surface of the earth, & found them answer pretty nearly. All this was in the two plague years of 1665 & 1666. For in those days I was in the prime of my age for invention & minded Mathematicks & Philosophy more then [sic] at any time since.[2]

More recently, the term *annus mirabilis* has been applied to the work of Albert Einstein during 1905 in an effort to establish a parallel between a crucial year in the life of the founding father of classical physics and of his twentieth-century successor.[3] What did Einstein accomplish during his miraculous year? We are fortunate in having his own contemporary summaries of his 1905 papers. Of the first four he wrote to a close friend:

I promise you four papers . . . , the first of which I could send you soon, since I will soon receive the free reprints. The paper deals with radiation and the energetic properties of light and is very revolutionary, as you will see. . . . The second paper is a determination of the true sizes of atoms from the diffusion and viscosity of dilute solutions of neutral substances. The third proves that, on the assumption of the molecular [kinetic] theory of heat, bodies of the order of magnitude of 1/1000 mm, suspended in liquids, must already perform an observable random movement that is produced by thermal motion; in fact, physiologists have observed motions of suspended small, inanimate, bodies, which they call "Brownian molecular motion." The fourth paper is only a rough draft at this point, and is an electrodynamics of moving bodies, which employs a modification of the theory of space and time; the purely kinematical part of this paper will surely interest you.[4]

Einstein characterized the fifth paper in these words:

One more consequence of the paper on electrodynamics has also occurred to me. The principle of relativity, in conjunction with Maxwell's equations, requires that mass be a direct measure of the energy contained in a body; light carries mass with it. A noticeable decrease of mass should occur in the case of radium. The argument is amusing and seductive; but for all I know, the Lord might be laughing over it and leading me around by the nose.[5]

The parallels are clear: each man was in his mid-twenties; each had given little previous sign of the incipient flowering of his genius; and, during a brief time span, each struck out on new paths that would ultimately revolutionize the science of his times. If Newton was only twenty-four in 1666 while Einstein was twenty-six in 1905, no one expects such parallels to be perfect.

While these parallels cannot be denied, upon closer inspection we can also see differences—much more significant than the slight disparity in age—between the activities of the two men during their *anni mirabiles* and in the immediate consequences of their work. The first striking difference is the one between their life situations: rejected by the academic community after graduation from the Swiss Polytechnical School in 1900, by 1905 Einstein was already a married man and an active father of a one-year-old son, obliged to fulfill the demanding responsibilities of a full-time job at the Swiss Patent Office. Newton never married (there is speculation that he died a virgin), and he had just taken his bachelor's degree but was still what we would call a graduate student in 1666. Indeed, he had been temporarily freed of even his academic responsibilities by the closure of Cambridge University after outbreaks of the plague.

Next we may note the difference in their scientific standing. Newton had published nothing by 1666, while Einstein already had published five respectable if not extraordinary papers in the prestigious *Annalen der Physik*. Thus, if 1666 marks the year when Newton's genius caught fire and he embarked on independent research, 1905 marks the year when Einstein's already matured talents manifested themselves to the world in a burst of creativity, a series of epoch-making works, all of which were published by the *Annalen* either in that year or the next. None of Newton's activities in 1666 found their way into print until much later: "The first blossoms of his genius flowered in private, observed silently by his own eyes alone in the years 1664 to 1666, his *anni mirabiles.*"[6] The reasons for Newton's evident lack of a need for recognition—indeed, his pronounced reluctance to share his ideas with others, as his major works had to be pried from his hands by others—have long been the topic of psychological, even psychopathological, speculation.

It took a few years—an agonizingly long time for a young man eager for recognition (see p. 115 below)—for Einstein's achievements to be fully acknowledged by the physics community. But the process started almost immediately in 1905; by 1909 Einstein had been called to a chair of theoretical physics created for him at the University of Zurich, and he was invited to lecture at the annual meeting of the assembled German-speaking scientific community.

Thus, if 1905 marks the beginning of the emergence of Einstein as a leading figure in the physics community, Newton remained in self-imposed obscurity well after 1666. Only in 1669, when at the urging of friends he allowed the limited circulation of a mathematical manuscript divulging some parts of the calculus he had developed, did "Newton's anonymity begin to dissolve."[7]

Another striking difference between the two is in their mathematical talents. Newton manifested his mathematical creativity from the outset. "In roughly a year [1664], without the benefit of instruction, he mastered the entire achievement of seventeenth-century analysis and began to break new ground. . . . The fact that he was unknown does not alter the fact that the young man not yet twenty-four, without benefit of formal instruction, had become the leading mathematician of Europe."[8]

Newton was thus able to create the mathematics necessary to develop his ideas about mechanics and gravitation.

Einstein, while an able pupil and practitioner, was never really creative in mathematics. Writing about his student years, Einstein said:

> The fact that I neglected mathematics to a certain extent had its cause not merely in my stronger interest in the natural sciences than in mathematics but also in the following peculiar experience. I saw that mathematics was split up into numerous specialties, each of which could easily absorb the short lifetime granted to us. Consequently, I saw myself in the position of Buridan's ass, which was unable to decide upon any particular bundle of hay. Presumably this was because my intuition was not strong enough in the field of mathematics to differentiate clearly the fundamentally important, that which is really basic, from the rest of the more or less dispensable erudition. Also, my interest in the study of nature was no doubt stronger; and it was not clear to me as a young student that access to a more profound knowledge of the more basic principles of physics depends on the most intricate mathematical methods. This dawned upon me only gradually after years of independent scientific work.[9]

Fortunately, for his works of 1905 he needed no more mathematics than he had been taught at school. Even so, it was left to Henri Poincaré, Hermann Minkowski, and Arnold Sommerfeld to give the special theory of relativity its most appropriate mathematical formulation.

When a really crucial need for new mathematics manifested itself in the course of his work on the general theory of relativity, Einstein had to make do with the tensor calculus as developed by Gregorio Ricci-Curbastro and Tullio Levi-Civita and presented to Einstein by his friend and colleague, Marcel Grossmann. This was based on Riemannian geometry, which lacked the concepts of parallel displacement and affine connection that would have so facilitated Einstein's work. But he was incapable of filling this mathematical lacuna, a task that was accomplished by Levi-Civita and Hermann Weyl only after the completion of the general theory.

Returning to Newton: in some respects he was right to hesitate about publication in 1666. "When 1666 closed, Newton was not in command of the results that have made his reputation deathless, not in mathematics, not in mechanics, not in optics. What he had done in all three was to lay foundations, some more extensive than others, on which he could build with assurance, but nothing was complete at the end of 1666, and most were not even close to complete."[10]

His work on the method of fluxions (as he called the calculus), even if incomplete, was worthy of publication and would have been of great service to contemporary mathematicians had it been available to them. His work in physics was far less advanced. His experiments on the theory of colors were interrupted by the closing of the university, and after his return to Cambridge in 1667 he spent a decade pursuing his optical investigations. Nevertheless, a more outgoing man might have published a preliminary account of his theory of colors in 1666. But in the case of gravitation, after carefully reviewing the evidence bearing on Newton's work on this subject through 1666, the physicist Leon Rosenfeld concluded that "it will be clear to every scientist that Newton at this stage had opened up for himself an exciting prospect, but had nothing fit to be published."[11] It is also clear that, in thinking about mechanics, he had not yet arrived at a clear concept of force—an essential prerequisite for the development of what we now call Newtonian mechanics. He had given "a new definition of force in which a body was treated as the passive subject of external forces impressed upon it instead of the active vehicle of forces impinging on others." But: "More than twenty years of patient if intermittent thought would in the end elicit his whole dynamics from this initial insight."[12]

. To sum up, in the case of Newton, in 1666 we have a student, working at his leisure, a mature genius in mathematics, but whose work in physics, however genial, was still in its formative stages. In the case of Einstein, in 1905 we have a man raising a family and pursuing a practical career, forced to fit physics into the interstices of an already-full life, yet already a master of theoretical physics ready to demonstrate that mastery to the world.

II

Newton's great legacy was his advancement of what at the time was called the mechanical philosophy and later came to be called the mechanical worldview. In physics, it was embodied in the so-called central force program: matter was assumed to be made up of particles of different species, referred to as "molecules." Two such molecules exerted various forces on each other: gravitational, electrical, magnetic, capillary, etc. These forces—attractive or repulsive—were assumed to be central, that is, to act in the direction of the line connecting the two particles, and to obey appropriate laws (such as the inverse square law for the gravitational and electrostatic forces), which depended on the distance between them. All physical phenomena were assumed to be explicable on the basis of Newton's three laws of motion applied to molecules acted upon by such central forces.

The central force program was shaken around the middle of the nineteenth century when it appeared that, in order to explain electromagnetic interactions between moving charged molecules, velocity- and acceleration-dependent forces had to be assumed. But it received the coup de grâce when Michael Faraday and James Clerk Maxwell's concept of the electromagnetic field began to prevail. According to the field point of view, two charged particles do not interact directly: each charge creates fields in the space surrounding it, and it is these fields which exert forces on the other charge. At first, these electric and magnetic fields were conceived of as states of a mechanical medium, the electromagnetic ether; these states were assumed ultimately to be explainable on the basis of mechanical models of that ether. Meanwhile, Maxwell's equations gave a complete description of the possible states of the electric and magnetic fields at all points of space and how they change over time. By the turn of the century, the search for mechanical explanations of the ether had been largely abandoned in favor of Hendrik Antoon Lorentz's viewpoint, frankly dualistic: the electric and magnetic fields were accepted as fundamental states of the ether, governed by Maxwell's equations but not in need of further explanation. Charged particles, which Lorentz called electrons (others continued to call them molecules or ions), obeyed Newton's mechanical laws of motion under the influence of forces that include the electric and magnetic forces exerted by the ether; and in turn the charged particles created these fields by their presence in and motion through the ether.

I call Lorentz's outlook dualistic because he accepted the mechanical worldview as applied to his electrons but regarded the ether with its electric and magnetic fields as an additional, independent element of reality, not mechanically explicable. To those brought up on the doctrine of the essential unity of nature, especially popular in Germany since the time of Alexander von Humboldt, such a dualism was uncomfortable if not intolerable.

Indeed, it was not long before Wilhelm Wien and others suggested another possibility: perhaps the electromagnetic field is the really fundamental entity, and the behavior of matter depends entirely on its electromagnetic properties. Instead of explaining the behavior of electromagnetic fields in terms of a mechanical model of the ether, this electromagnetic worldview hoped to explain the mechanical properties of matter in terms of electric and magnetic fields. Even Lorentz flirted with this possibility, though he never fully adopted it.

The mechanical worldview did not simply disappear with the advent of Maxwell's electrodynamics. The last third of the nineteenth century saw a remarkable new triumph of the mechanical program. On the basis of the application of statistical methods to large assemblies of molecules (Avogadro's number, about 6.3×10^{23} molecules per mole of any substance, here gives the measure of largeness), Maxwell and Ludwig Boltzmann succeeded in giving a mechanical foundation to the laws of thermodynamics and started the program of explaining the bulk properties of matter in terms of kinetic-molecular theories of the gaseous, liquid, and solid states.

III

Thus, as a student Einstein had to master both the traditional mechanical viewpoint, particularly its application to the atomistic picture of matter, as well as Maxwell's new field-theoretical approach to electromagnetism, particularly in Lorentz's version. He was also confronted with a number of new phenomena, such as black-body radiation and the photoelectric effect, which stubbornly resisted all attempts to fit them into either the old mechanical or the new electromagnetic worldview—or any combination of the two. From this perspective, his five epoch-making papers of 1905 may be divided into three categories. The first two categories concern extensions and modifications of the two physical theories that dominated physics at the end of the nineteenth century: classical mechanics and Maxwell's electrodynamics.

1. His two papers on molecular dimensions and Brownian motion, papers 1 and 2 in this volume, are efforts to extend and perfect the classical-mechanical approach, especially its kinetic-molecular implications.

2. His two papers on the theory of special relativity, papers 3 and 4, are efforts to extend and perfect Maxwell's theory by modifying the foundations of classical mechanics in order to remove the apparent contradiction between mechanics and electrodynamics.

In these four papers, Einstein proved himself a master of what we today call classical physics, the inheritor and continuer of the tradition that started with Galileo Galilei and Newton and ended with Faraday, Maxwell, and Boltzmann, to name but a few of the most outstanding representatives of this tradition. Revolutionary as they then appeared to his contemporaries, the new insights into the nature of space, time, and motion necessary to develop the special theory of relativity are now seen as the climax and culmination of that classical tradition.

3. His work on the light quantum hypotheses, paper 5, is the only one that he himself regarded as truly radical. In the first letter cited on p. 5 above, he wrote that this paper "deals with radiation and the energetic properties of light and is very revolutionary."[13] In it, he demonstrated the limited ability of both classical mechanics and Maxwell's electromagnetic theory to explain the properties of electromagnetic radiation, and introduced the hypothesis that light has a granular structure in order to explain novel phenomena such as the photoelectric effect, which cannot be explained on the basis of classical physics. Here and subsequently, Einstein, master of the classical tradition, proved to be its most severe and consistent critic and a pioneer in the search to find a new unified foundation for all of physics.

IV

The papers are presented in this volume in the order suggested by the three categories mentioned above, roughly the order of their distance from classical physics; but the reader should feel no compulsion to read them in that order. A good case can be made for the chronological order, for jumping immediately to the papers on special relativity and quantum theory—or for simply dipping into the volume as one's interest or fancy dictates.

In the body of this volume, the reader will find detailed discussions of each of these five papers drawn from the thematic introductory essays in volume 2 of *The Collected*

Papers of Albert Einstein. Here I shall give an overview of Einstein's work up to and including 1905 in each of the three categories.

1. *Efforts to Extend and Perfect the Classical-Mechanical Tradition*

As recently discovered letters show, by the turn of the century Einstein was already occupied with the problems that were to take him beyond classical physics. Yet all of his papers published before 1905 treat topics that fall within the framework of Newtonian mechanics and its applications to the kinetic-molecular theory of matter. In his first two papers, published in 1901 and 1902, Einstein attempted to explain several apparently quite different phenomena occurring in liquids and solutions on the basis of a single simple hypothesis about the nature of the central force between molecules, and how it varies with their chemical composition. Einstein hoped that his work might help to settle the status of a long-standing (and now discarded) conjecture about a common basis for molecular and gravitational forces—one indication of his strong ambition from the outset to contribute to the theoretical unification of all the apparently disparate phenomena of physics. In 1901 he wrote: "It is a wonderful feeling to realize the unity of a complex of phenomena which, to immediate sensory perception, appear to be totally separate things."[14] Much later, looking back over his life, he wrote: "The real goal of my research has always been the simplification and unification of the system of theoretical physics."[15]

As mentioned on p. 12, another great project of nineteenth-century physics was the attempt to show that the empirically well-verified laws of thermodynamics could be explained theoretically on the basis of an atomistic model of matter. Maxwell and Boltzmann were pioneers in this effort, and Einstein saw himself as continuing and perfecting their work.

Einstein made extensive use of thermodynamical arguments in his first two papers; indeed, thermodynamics plays an important role in all of his early work. The second paper raises a question about the relation between the thermodynamic and kinetic-molecular approaches to thermal phenomena that he answered in his next paper. This is the first of three, published between 1902 and 1904, devoted to the atomistic foundations of thermodynamics. His aim was to formulate the minimal atomistic assumptions about a mechanical system needed to derive the basic concepts and principles of thermodynamics. Presumably because he derived it from such general assumptions, he regarded the second law of thermodynamics as a "necessary consequence of the mechanical worldview."[16] He also derived an equation for the mean square energy fluctuations of a system in thermal equilibrium. In spite of its mechanical origins, this formula involves only thermodynamical quantities, and Einstein boldly proceeded to apply the equation to an apparently nonmechanical system: black-body radiation (his first mention of it in print), that is, electromagnetic radiation in thermal equilibrium with matter. Black-body radiation was the only system for which it was clear to him that energy fluctuations should be physically significant on an observable length scale, and his calculations proved consistent with the known properties of that radiation. This calculation suggests that Einstein may already have had in mind an attempt to treat black-body radiation as if it were a mechanical system—the basis of his "very revolutionary" light quantum hypothesis of 1905.

In paper 1 of this volume, his doctoral dissertation, Einstein used methods based on classical hydrodynamics and diffusion theory to show that measurement of a fluid's viscosity with and without the presence of a dissolved substance can be used to obtain an estimate of Avogadro's number (see p. 12) and the size of the molecules of the dissolved substance. Paper 2, the so-called Brownian-motion paper, also extends the scope of applicability of classical mechanical concepts. Einstein noted that, if the kinetic-molecular theory of heat is correct, the laws of thermodynamics cannot be universally valid, since fluctuations must give rise to microscopic but visible violations of the second law when one considers particles sufficiently large for their motion to be observable in a microscope if suspended in a liquid. Indeed, as Einstein showed, such fluctuations explain the well-known Brownian motion of microscopic particles suspended in a liquid. He regarded his work as establishing the limits of validity within which thermodynamics could be applied with complete confidence.

2. Efforts to Extend and Perfect Maxwell's Electrodynamics and Modify Classical Mechanics to Cohere with It

Well before 1905, Einstein apparently was aware of a number of experiments suggesting that the mechanical principle of relativity—the equivalence of all inertial frames of reference for the description of any mechanical phenomena—should be extended from mechanical to optical and electromagnetic phenomena. However, such an extension was in conflict with what he regarded as the best current electrodynamical theory, Lorentz's electron theory, which grants a privileged status to one inertial frame: the ether rest frame (see p. 11).

In papers 3 and 4 in this volume, Einstein succeeded in resolving this conflict through a critical analysis of the kinematical foundations of physics, the theory of space and time, which underlies mechanics, electrodynamics, and indeed (although no others were known at the time) any other dynamical theory. After a profound critical study of the concept of simultaneity of distant events, Einstein realized that the principle of relativity could be made compatible with Maxwell's equations if one abandoned Newtonian absolute time in favor of a new absolute: the speed of light, the same in all inertial frames. As a consequence, the Newtonian-Galileian laws of transformation between the space and time coordinates of different inertial frames must be replaced by a set of transformations, now called the Lorentz transformations.[17] Since these transformations are kinematical in nature, any acceptable physical theory must be invariant under the group of such transformations. Maxwell's equations, suitably reinterpreted after eliminating the concept of the ether, meet this requirement; but Newton's equations of motion needed revision.

Einstein's work on the theory of relativity provides an example of his ability to move forward amid paradox and contradiction. He employs one theory—Maxwell's electrodynamics—to find the limits of validity of another—Newtonian mechanics—even though he was already aware of the limited validity of the former (see pp. 20–22 below).

One of the major accomplishments of Einstein's approach, which his contemporaries found difficult to apprehend, is that relativistic kinematics is independent of the theories that impelled its formulation. He had not only formulated a coherent kinematical basis for both mechanics and electrodynamics, but (leaving aside the problem of gravitation) for any new physical concepts that might be introduced. Indeed, developments in physics over almost a century have not shaken these kinematical foundations. To use terms that he employed later, Einstein had created a theory of principle, rather than a constructive theory.[18] At the time he expressed the distinction in these words: "One is in no way dealing here . . . with a 'system' in which the individual laws would implicitly be contained and could be found merely by deduction therefrom, but only with a principle that (in a way similar to the second law of thermodynamics) permits the reduction of certain laws to others."[19] The principles of such a theory, of which thermodynamics is his prime example, are generalizations drawn from a large amount of empirical data that they summarize and generalize without purporting to explain. In contrast, constructive theories, such as the kinetic theory of gases, do purport to explain certain phenomena on the basis of hypothetical entities, such as atoms in motion, introduced precisely to provide such explanations.

It is well known that important elements of Einstein's distinction between principle and constructive theories are found in Poincaré's writings. Two lesser-known sources that may have influenced Einstein's emphasis on the role of principles in physics are the writings of Julius Violle and Alfred Kleiner, which he is also known to have read.

In spite of the merits of the theory of relativity, however, Einstein felt that it was no substitute for a constructive theory: "A physical theory can be satisfactory only if its structures are composed of elementary foundations. The theory

19

of relativity is just as little ultimately satisfactory as, for example, classical thermodynamics was before Boltzmann had interpreted the entropy as probability."[20]

3. Demonstrations of the Limited Validity of Both Classical Mechanics and Maxwell's Electromagnetic Theory, and Attempts to Comprehend Phenomena That Cannot Be Explained by These Theories

Einstein's efforts to perfect classical mechanics and Maxwell's electrodynamics, and to make both theories compatible, may still be regarded as extensions, in the broadest sense, of the classical approach to physics. However original his contributions in these areas may have been, however revolutionary his conclusions about space and time appeared to his contemporaries, however fruitful his work proved to be for the exploration of new areas of physics, he was still engaged in drawing the ultimate consequences from conceptual structures that were well established by the end of the nineteenth century. What is unique about his stance during the first decade of this century is his unwavering conviction that classical mechanical concepts and those of Maxwell's electrodynamics—as well as any mere modification or supplementation of the two—are incapable of explaining a growing list of newly discovered phenomena involving the behavior and interactions of matter and radiation. Einstein constantly reminded his colleagues of the need to introduce radically new concepts to explain the structure of both matter and radiation. He himself introduced some of these new concepts, notably the light quantum hypothesis, although he remained unable to integrate them into a coherent physical theory.

Paper 5, Einstein's first paper on the quantum hypothesis, is a striking example of his style, mingling critique of old concepts with the search for new ones. It opens by demonstrating that the equipartition theorem,[21] together with Maxwell's equations, leads to a definite formula for the black-body radiation spectrum, now known as the Rayleigh-Jeans distribution. This distribution, which at low frequencies matches the empirically validated Planck distribution, cannot possibly hold at high frequencies, since it implies a divergent total energy. (He soon gave a similar demonstration, also based on the equipartition theorem, that classical mechanics cannot explain the thermal or optical properties of a solid, modeled as a lattice of atomic or ionic oscillators.)

Einstein next investigated this high-frequency region, where the classically derived distribution breaks down most dramatically. In this region, called the Wien limit, he showed that the entropy of monochromatic radiation with a fixed temperature depends on its volume in exactly the same way as does the entropy of an ordinary gas composed of statistically independent particles. In short, monochromatic radiation in the Wien limit behaves thermodynamically as if it were composed of statistically independent quanta of energy. To obtain this result, Einstein had to assume each quantum has an energy proportional to its frequency. Emboldened by this result, he took the final step, proposing his "very revolutionary" hypothesis that matter and radiation can interact only through the exchange of such energy quanta. He demonstrated that this hypothesis explains a number of apparently disparate phenomena, notably the photoelectric effect; it was this work that was cited by the Nobel Prize committee in 1921.

In 1905 Einstein did not use Planck's full distribution law. The following year he showed that Planck's derivation of this law implicitly depends on the assumption that the energy of charged oscillators can only be an integral multiple of the quantum of energy, and hence these oscillators can only exchange energy with the radiation field by means of such quanta. In 1907, Einstein argued that uncharged oscillators should be similarly quantized, thereby explaining both the success of the DuLong-Petit law for most solids at ordinary temperatures and the anomalously low values of the specific heats of certain substances. He related the temperature at which departures from the DuLong-Petit law (see p. 175) become significant—now called the Einstein temperature— to the fundamental frequency of the atomic oscillators, and hence to the optical absorption spectrum of a solid.

In spite of his conviction of its fundamental inadequacy, Einstein continued to utilize still-reliable aspects of classical mechanics with remarkable skill to explore the structure of electromagnetic radiation. In 1909 he applied his theory of Brownian motion to a two-sided mirror immersed in thermal radiation. He showed that the mirror would be unable to carry out such a Brownian motion indefinitely if the fluctuations of the radiation pressure on its surfaces were due solely to the effects of random waves, as predicted by Maxwell's theory. Only the existence of an additional term, corresponding to pressure fluctuations due to the impact of random particles on the mirror, guarantees its continued Brownian motion. Einstein showed that both wave and particle energy fluctuation terms are consequences of Planck's distribution law for black-body radiation. He regarded this result as his strongest argument for ascribing physical reality to light quanta.

Einstein was far from considering his work on the quantum hypothesis as constituting a satisfactory theory of radiation or matter. As noted on p. 19, he emphasized that a physical theory is satisfactory only "if its structures are composed of *elementary* foundations," adding "that we are still far from having satisfactory elementary foundations for electrical and mechanical processes."[22] Einstein felt that he had not achieved a real understanding of quantum phenomena because (in contrast to his satisfactory interpretation of Boltzmann's constant as setting the scale of statistical fluctuations) he had been unable to interpret Planck's constant "in an intuitive way."[23] The quantum of electric charge also remained "a stranger" to theory.[24] He was convinced that a satisfactory theory of matter and radiation must construct these quanta of electricity and of radiation, not simply postulate them.

As a theory of principle (see above), the theory of relativity provides important guidelines in the search for such a satisfactory theory. Einstein anticipated the ultimate construction of "a complete worldview that is in accord with the principle of relativity."[25] In the meantime, the theory offered clues to the construction of such a worldview. One clue concerns the structure of electromagnetic radiation. Not only is the theory compatible with an emission theory of radiation, since it implies that the velocity of light is always the same relative to its source; the theory also requires that radiation transfer mass between an emitter and an absorber, reinforcing Einstein's light quantum hypothesis that radiation manifests a particulate structure under certain circumstances. He maintained that "the next phase in the development of theoretical physics will bring us a theory of light, which may be regarded as a sort of fusion of the undulatory and emission theories of light."[26] Other principles that Einstein regarded as reliable guides in the search for an understanding of quantum phenomena are conservation of energy and Boltzmann's principle.

Einstein anticipated that "the same theoretical modification that leads to the elementary quantum [of charge] will also lead to the quantum structure of radiation as a consequence."[27] In 1909 he made his first attempt to find a field theory that would explain both the structure of matter (the electron) and of radiation (the light quantum). After investigating relativistically invariant, non-linear generalizations of Maxwell's equations, he wrote: "I have not succeeded . . . in finding a system of equations that I could see was suited to the construction of the elementary quantum of electricity and the light quantum. The manifold of possibilities does not seem to be so large, however, that one need draw back in fright from the task."[28] This attempt may be regarded as the forerunner of his later, almost forty-year-long search for a unified field theory of electromagnetism, gravitation, and matter.

In 1907, Einstein's attempt to incorporate gravitation into the theory of relativity led him to recognize a new formal principle, the principle of equivalence, which he interpreted as demonstrating the need to generalize the relativity principle (which he now began to call the *special* relativity principle) if gravitation is to be included in its scope. He found that, when gravitational effects are taken into account, it is impossible to maintain the privileged role that inertial frames of reference and Lorentz transformations play in the original relativity theory. He started the search for a group of transformations wider than the Lorentz group, under which the laws of physics remain invariant when gravitation is taken into account. This search, which lasted until the end of 1915, culminated in what Einstein considered his greatest scientific achievement: the general theory of relativity—but that is another story, which I cannot tell here.

Nor can I do more than allude to the many ways in which Einstein's work on the special theory of relativity and the quantum theory have inspired and guided not only many of the revolutionary transformations of our picture of the physical world during the twentieth century, but—through their influence on technological development—have contributed to equally revolutionary transformations in our way of life. One cannot mention quantum optics or quantum field theory, to name only a couple of theoretical advances; nor masers and lasers, klystrons and synchrotrons—nor atomic and hydrogen bombs, to name only a few of the multitude of inventions that have changed our world for good or ill, without invoking the heritage of Einstein's miraculous year.

EDITORIAL NOTES

[1]The phrase *anni mirabiles* (years of wonders) has been applied with more accuracy to the years 1664–1666 by Newton's biographer Richard Westfall in *Never at Rest/A Biography of Isaac Newton* (Cambridge, U.K.: Cambridge University Press, 1980; paperback edition, 1983), p. 140. This book may be consulted for generally reliable biographical information about Newton's life.

[2]I. Bernard Cohen, *Introduction to Newton's 'Principia'* (Cambridge, Mass.: Harvard University Press, 1971), p. 291.

[3]See, for example, Albrecht Fölsing, *Albert Einstein/A Biography*, tr. by Ewald Osers (New York: Viking, 1997), p. 121: "Never before and never since has a single person enriched science by so much in such a short time as Einstein did in his *annus mirabilis*." This book may be consulted for generally reliable biographical information about Einstein, but its scientific explanations should be treated with caution. For an account of Einstein's scientific work organized biographically, see Abraham Pais,

'Subtle is the Lord . . . ': The Science and the Life of Albert Einstein (Oxford: Clarendon Press; New York: Oxford University Press, 1982).

[4] Einstein to Conrad Habicht, 18 or 25 May 1905, The Collected Papers of Albert Einstein (Princeton, N.J.: Princeton University Press, 1987–), cited hereafter as Collected Papers, vol. 5 (1993), doc. 27, p. 31. Translation from Anna Beck, tr., The Collected Papers of Albert Einstein: English Translation (Princeton University Press, 1987–), cited hereafter as English Translation, vol. 5 (1995), p. 20; translation modified.

[5] Einstein to Conrad Habicht, 30 June–22 September 1905, Collected Papers, vol. 5, doc. 28, p. 33; English Translation, p. 21; translation modified. Forty years later, when the explosion of the first atomic bombs brought the equivalence between mass and energy forcefully to the world's attention, Einstein might have wondered just what sort of trick the Lord had played on him.

[6] Westfall, Never at Rest, p. 140.

[7] Ibid., p. 205.

[8] Ibid., pp. 100, 137.

[9] Albert Einstein, Autobiographical Notes, Paul Arthur Schilpp, ed. and trans. (LaSalle, Ill.: Open Court, 1979), p. 15.

[10] Westfall, Never at Rest, p. 174.

[11] "Newton and the Law of Gravitation," Arch. Hist. Exact Sci. 2 (1965): 365–386, reprinted in Robert S. Cohen and John J. Stachel, eds., Selected Papers of Leon Rosenfeld (Dordrecht/Boston: Reidel, 1979), p. 65.

[12] Citation from Westfall, Never at Rest, p. 146.

[13] Einstein to Conrad Habicht, May 1905, Collected Papers, vol. 5, doc. 27, p. 31.

[14] Einstein to Marcel Grossmann, 14 April 1901, Collected Papers, vol. 1, doc. 100, p. 290.

[15] From the response to a questionnaire submitted to Einstein in 1932. See Helen Dukas and Banesh Hoffmann, Albert Einstein: The Human Side (Princeton, N.J.: Princeton University Press, 1979), p. 11 for the English translation, p. 122 for the German text.

[16] Einstein, "Kinetic Theory of Thermal Equilibrium and of the Second Law of Thermodynamics," in Collected Papers, vol. 2, doc. 3, p. 72 (p. 432 of the 1902 original).

[17] Lorentz had introduced such a set of transformations, and Henri Poincaré had so named them; but the kinematical interpretation that Einstein gave to them is quite different.

[18] For the distinction between theories of principle and constructive theories, see Albert Einstein, "Time, Space and Gravitation," The Times (London), 28 November 1919, p. 13; reprinted as "What Is the Theory of

Relativity?" in *Ideas and Opinions* (New York: Crown, 1954), pp. 227–232. He later reminisced about the origins of the theory: "Gradually I despaired of the possibility of discovering the true laws by means of constructive efforts based on known facts. The longer and the more desperately I tried, the more I came to the conviction that only the discovery of a universal formal principle could lead us to assured results. The example I saw before me was thermodynamics" (*Autobiographical Notes*, p. 48; translation, p. 49). For several years after 1905, Einstein referred to the "relativity principle" rather than to the "theory of relativity."

[19] Einstein, "Comments on the Note of Mr. Paul Ehrenfest: 'The Translatory Motion of Deformable Electrons and the Area Law,'" in *Collected Papers*, vol. 2, doc. 44, p. 411 (p. 207 of the 1907 original).

[20] Einstein to Arnold Sommerfeld, 14 January 1908, *Collected Papers*, vol. 5, doc. 73, pp. 86–88. A decade later, Einstein elaborated this idea: "When we say that we have succeeded in understanding a group of natural processes, we always mean by this that a constructive theory has been found, which embraces the processes in question" (from "Time, Space and Gravitation").

[21] This is a result of classical statistical mechanics, according to which each degree of freedom of a mechanical system in thermal equilibrium receives, on the average, the same share of the total energy of the system.

[22] Einstein to Arnold Sommerfeld, 14 January 1908, *Collected Papers*, vol. 5, doc. 73, p. 87.

[23] Ibid.

[24] See Einstein, "On the Present Status of the Radiation Problem," in *Collected Papers*, vol. 2, doc. 56, p. 549 (p. 192 of the 1909 original).

[25] Einstein, "On the Inertia of Energy Required by the Relativity Principle," in *Collected Papers*, vol. 2, doc. 45, pp. 414–415 (p. 372 of the 1907 original).

[26] Einstein, "On the Development of Our Views Concerning the Nature and Constitution of Radiation," in *Collected Papers*, vol. 2, doc. 60, pp. 564–565 (pp. 482–483 of the 1909 original).

[27] Einstein, "On the Present Status of the Radiation Problem," in *Collected Papers*, vol. 2, doc. 56, pp. 549–550 (pp. 192–193 of the 1909 original).

[28] Ibid., p. 550 (p. 193 of the 1909 original). This attempt at a field theory seems to represent Einstein's first step toward a field ontology.

Part One

✦

*Einstein's Dissertation on
the Determination of
Molecular Dimensions*

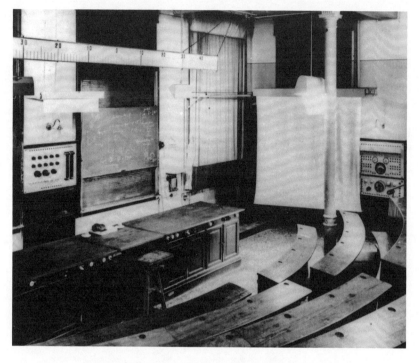

Lecture hall in the Physics Building, Eidgenössische
Technische Hochschule, Zurich, 1905. (Courtesy of ETH)

Einstein submitted a dissertation to the University of Zurich in 1901, about a year after graduation from the Eidgenössische Technische Hochschule (ETH), but withdrew it early in 1902. In a successful second attempt three years later, he combined the techniques of classical hydrodynamics with those of the theory of diffusion to create a new method for the determination of molecular sizes and of Avogadro's number, a method he applied to solute sugar molecules. The dissertation was completed on 30 April 1905 and submitted to the University of Zurich on 20 July. On 19 August 1905, shortly after the thesis was accepted, the *Annalen der Physik* received a slightly different version for publication.

By 1905, several methods for the experimental determination of molecular dimensions were available. Although estimates of upper bounds for the sizes of microscopic constituents of matter had been discussed for a long time, the first reliable methods for determining molecular sizes were developed in the second half of the nineteenth century, based on the kinetic theory of gases. The study of phenomena as diverse as contact electricity in metals, the dispersion of light, and black-body radiation yielded new approaches to the problem of molecular dimensions. Most of the methods available by the turn of the century gave values for the size of molecules and for Avogadro's number that are in more or less satisfactory agreement with each other.

Although Einstein claimed that the method in his dissertation is the first to use phenomena in fluids in the determination of molecular dimensions, the behavior of liquids plays a role in various earlier methods. For example, the comparison of densities in the liquid and gaseous states is an important part of Loschmidt's method, based on the kinetic theory of gases. A method that depends entirely on the physics of liquids was developed as early as 1816 by Thomas Young. Young's study of surface tension in liquids led to an estimate of the range of molecular forces, and capillary phenomena were used later in several different ways to determine molecular sizes.

A kinetic theory of liquids, comparable to the kinetic theory of gases, was not available, and the methods for deriving molecular volumes exclusively from the properties of liquids did not give very precise results. Einstein's method, on the other hand, yields values comparable in precision to those provided by the kinetic theory of gases. While methods based on capillarity presuppose the existence of molecular forces, Einstein's central assumption is the validity of using classical hydrodynamics to calculate the effect of solute molecules, treated as rigid spheres, on the viscosity of the solvent in a dilute solution.

Einstein's method is well suited to determine the size of solute molecules that are large compared to those of the solvent. In 1905 William Sutherland published a new method for determining the masses of large molecules that shares important elements with Einstein's. Both methods make use of the molecular theory of diffusion that Nernst developed on the basis of van't Hoff's analogy between solutions and gases, and of Stokes's law of hydrodynamical friction.

Sutherland was interested in the masses of large molecules because of the role they play in the chemical analysis of organic substances such as albumin. In developing a new method for the determination of molecular dimensions, Einstein was concerned with several other problems on different levels of generality. An outstanding current problem of the theory of solutions was whether molecules of the solvent are attached to the molecules or ions of the solute. Einstein's dissertation contributed to the solution of this problem. He recalled in a letter to Jean Perrin in November 1909: "At the time I used the viscosity of the solution to determine the volume of sugar dissolved in water because in this way I hoped to take into account the volume of any *attached* water molecules." The results obtained in his dissertation indicate that such an attachment does occur.

Einstein's concerns extended beyond this particular question to more general problems of the foundations of the theory of radiation and the existence of atoms. He later emphasized in the same letter: "A precise determination of the size of molecules seems to me of the highest importance because Planck's radiation formula can be tested more precisely through such a determination than through measurements on radiation."

The dissertation also marked the first major success in Einstein's effort to find further evidence for the atomic hypothesis, an effort that culminated in his explanation of Brownian motion. By the end of 1905 he had published three independent methods for determining molecular dimensions, and in the following years he found several more. Of all these methods, the one in his dissertation is most closely related to his earlier studies of physical phenomena in liquids.

EINSTEIN'S efforts to obtain a doctoral degree illuminate some of the institutional constraints on the development of his work on the problem of molecular dimensions. His choice of a theoretical topic for a dissertation at the University of Zurich was quite unusual, both because it was theoretical and because a dissertation theme was customarily assigned by the supervising professor. By 1900, theoretical physics was slowly beginning to achieve recognition as an independent discipline in German-speaking countries, but it was not yet established at either the ETH or the University of Zurich. A beginning had been made at the ETH soon after its founding, with the appointment of a German mathematical physicist, Rudolf Clausius. His departure a decade later may have been hastened by lack of official sympathy for a too-theoretical approach to the training of engineers and secondary-school teachers, the primary task of the school.

Clausius's successor—after the position had been vacant for a number of years—was H. F. Weber, who occupied the chair for Mathematical and Technical Physics from 1875 until his death in 1912. During the last two decades of the nineteenth century, he did original research, mainly in experimental physics and electrotechnology, including work on a number of topics that were important for Einstein's later research, such as black-body radiation, the anomalous low-temperature behavior of specific heats, and the theory of diffusion; but his primary interests were never those of a theoretical physicist. The situation of theoretical physics at the University of Zurich at the turn of the century was hardly better. Four other major Swiss universities either had two full professorships in physics or one full and one nontenured position, while Zurich had only one physics chair, held by the experimentalist Alfred Kleiner.

Since the ETH was not authorized to grant doctoral degrees until 1909, a special arrangement enabled ETH students to obtain doctorates from the University of Zurich. Most dissertations in physics by ETH students were prepared under Weber's supervision, with Kleiner as the second referee. As noted above, almost all physics dissertations prepared at the ETH and the University of Zurich between 1901 and 1905 were on experimental topics suggested to the students by their supervisor or at least closely related to the latter's research interests. The range of topics was quite limited, and generally not at the forefront of experimental research. Thermal and electrical conductivity, and instruments for their measurement, were by far the most prominent subjects. General questions of theoretical physics, such as the properties of the ether or the kinetic theory of gases, occasionally found their way into examination papers, but they were hardly touched upon in dissertations.

In the winter semester of 1900–1901, Einstein intended to work for a degree under Weber. The topic may have been related to thermoelectricity, a field in which Einstein had shown an interest and in which several of Weber's doctoral students did experimental research. After a falling-out with Weber, Einstein turned to Kleiner for advice and comments on his work.

Although Kleiner's research at this time focused on measuring instruments, he did have an interest in foundational questions of physics, and Einstein's discussions with him covered a wide range of topics. Einstein showed his first dissertation to Kleiner before submitting it to the university in November 1901. This dissertation has not survived, and the evidence concerning its contents is somewhat ambiguous. In April 1901 Einstein wrote that he planned to summarize his work on molecular forces, up to that time mainly on liquids; at the end of the year, his future wife Mileva Marić stated that he had submitted a work on molecular forces in gases. Einstein himself wrote that it concerned "a topic in the

kinetic theory of gases." There are indications that the dissertation may have discussed Boltzmann's work on gas theory,
as well as Drude's work on the electron theory of metals.

By February 1902 Einstein had withdrawn the dissertation, possibly at Kleiner's suggestion that he avoid a controversy with Boltzmann. In view of the predominantly experimental character of the physics dissertations submitted to
the University of Zurich at the time, lack of experimental
confirmation for his theoretical results may have played a
role in the decision to withdraw the thesis. In January 1903
Einstein still expressed interest in molecular forces, but he
stated in a letter to Michele Besso that he was giving up his
plan to obtain a doctorate, arguing that it would be of little help to him, and that "the whole comedy has become
tiresome for me."

Little is known about when Einstein started to work on
the dissertation he completed in 1905. By March 1903 some
of the central ideas of the 1905 dissertation had already occurred to him. Kleiner, one of the two faculty reviewers of
his dissertation, acknowledged in his review that Einstein
had chosen the topic himself and pointed out that "the arguments and calculations to be carried out are among the most
difficult in hydrodynamics." The other reviewer, Heinrich
Burkhardt, Professor of Mathematics at the University of
Zurich, added: "The mode of treatment demonstrates fundamental mastery of the relevant mathematical methods."
Although Burkhardt checked Einstein's calculations, he
overlooked a significant error in them. The only reported
criticism of Einstein's dissertation was for being too short.
Einstein's biographer Carl Seelig reports: "Einstein later
laughingly recounted that his dissertation was at first returned to him by Kleiner with the comment that it was too

short. After he had added a single sentence, it was accepted without further comment."

Compared to the other topics of his research at the time, his hydrodynamical method for determining molecular dimensions was a dissertation topic uniquely suited to the empirically oriented Zurich academic environment. In contrast to the Brownian-motion work, for which the experimental techniques needed to extract information from observations were not yet available, Einstein's hydrodynamical method for determining the dimensions of solute molecules enabled him to derive new empirical results from data in standard tables.

LIKE Loschmidt's method based on the kinetic theory of gases, Einstein's method depends on two equations for two unknowns, Avogadro's number N and the molecular radius P. The first of Einstein's equations (see third equation on p. 64) follows from a relation between the coefficients of viscosity of a liquid with and without suspended molecules (k^* and k, respectively),

$$k^* = k(1 + \varphi), \tag{1}$$

where φ is the fraction of the volume occupied by the solute molecules. This equation, in turn, is derived from a study of the dissipation of energy in the fluid.

Einstein's other fundamental equation follows from an expression for the coefficient of diffusion D of the solute. This expression is obtained from Stokes's law for a sphere of radius P moving in a liquid, and van't Hoff's law for the osmotic pressure:

$$D = \frac{RT}{6\pi k} \cdot \frac{1}{NP}, \tag{2}$$

where R is the gas constant, T the absolute temperature, and N Avogadro's number.

The derivation of eq. (1), technically the most complicated part of Einstein's thesis, presupposes that the motion of the fluid can be described by the hydrodynamical equations for stationary flow of an incompressible homogeneous liquid, even in the presence of solute molecules; that the inertia of these molecules can be neglected; that they do not affect each other's motions; and that they can be treated as rigid spheres moving in the fluid without slipping, under the sole influence of hydrodynamical stresses. The hydrodynamic techniques needed are derived from Kirchhoff's *Vorlesungen über mathematische Physik*, volume 1, *Mechanik* (1897), a book that Einstein first read during his student years.

Eq. (2) follows from the conditions for the dynamical and thermodynamical equilibrium of the fluid. Its derivation requires the identification of the force on a single molecule, which appears in Stokes's law, with the apparent force due to the osmotic pressure. The key to handling this problem is the introduction of fictitious countervailing forces. Einstein had earlier introduced such fictitious forces to counteract thermodynamical effects in proving the applicability to diffusion phenomena of a generalized form of the second law of thermodynamics, and in his papers on statistical physics.

Einstein's derivation of eq. (2) does not involve the theoretical tools he developed in his work on the statistical foundations of thermodynamics; he reserved a more elaborate derivation, using these methods, for his first paper on Brownian motion. Eq. (2) was derived independently, in somewhat more general form, by Sutherland in 1905. To deal with the available empirical data, Sutherland had to allow for a varying coefficient of sliding friction between the diffusing molecule and the solution.

The basic elements of Einstein's method—the use of diffusion theory and the application of hydrodynamical techniques to phenomena involving the atomistic constitution of matter or electricity—can be traced back to his earlier work. Einstein's previous work had touched upon most aspects of the physics of liquids in which their molecular structure is assumed to play a role, such as Laplace's theory of capillarity, Van der Waals's theory of liquids, and Nernst's theory of diffusion and electrolytic conduction.

Before Einstein's dissertation, the application of hydrodynamics to phenomena involving the atomic constitution of matter or electricity was restricted to consideration of the effects of hydrodynamical friction on the motion of ions. Stokes's law was employed in methods for the determination of the elementary charge and played a role in studies of electrolytic conduction. Einstein's interest in the theory of electrolytic conduction may have been decisive for the development of some of the main ideas in his dissertation. This interest may have suggested a study of molecular aggregates in combination with water, as well as some of the techniques used in the dissertation.

In 1903 Einstein and Besso discussed a theory of dissociation that required the assumption of such aggregates, the "hypothesis of ionic hydrates," as Besso called it, claiming that this assumption resolves difficulties with Ostwald's law of dilution. The assumption also opens the way to a simple calculation of the sizes of ions in solution, based on hydrodynamical considerations. In 1902 Sutherland had considered a calculation of the sizes of ions on the basis of Stokes's formula, but rejected it as in disagreement with experimental data. Sutherland did not use the assumption of ionic hydrates, which can avoid such disagreement by permitting ionic sizes to vary with such physical conditions as temperature and concentration. The idea of determining the sizes of ions by means of classical hydrodynamics occurred to Einstein in March 1903, when he proposed in a letter to Besso what appears to be just the calculation that Sutherland had rejected:

> Have you already calculated the absolute magnitude of ions on the assumption that they are spheres and so large that the hydrodynamical equations for viscous fluids are applicable? With our knowledge of the absolute magnitude of the electron [charge] this would be a simple matter indeed. I would have done it myself but lack the reference material and the time; you could also bring in diffusion in order to obtain information about neutral salt molecules in solution.

This passage is remarkable, because both key elements of Einstein's method for the determination of molecular dimensions, the theories of hydrodynamics and diffusion, are already mentioned, although the reference to hydrodynamics probably covers only Stokes's law. While a program very similar to the first of Einstein's proposals to Besso was being pursued at the time by William Robert Bousfield, Einstein's dissertation can be seen to be an elaboration of the second proposal, regarding diffusion and neutral salt molecules. Einstein may thus have been proceeding similarly to Nernst, who first developed his theory of diffusion for the simpler case of nonelectrolytes. The study of sugar solutions could draw upon extensive and relatively precise numerical data on viscosity and the diffusion coefficient, avoiding problems of dissociation and electrical interactions.

THE RESULTS obtained with Einstein's method for the determination of molecular dimensions differed from those obtained by other methods at the time, even when new data taken from Landolt and Bornstein's physical-chemical tables were used to recalculate them. In his papers on Brownian motion, Einstein cited either the value he obtained for Avogadro's number, or a more standard one. Only once, in 1908, did he comment on the uncertainty in the determination of this number. By 1909 Perrin's careful measurements of Brownian motion produced a new value for Avogadro's number, significantly different from the values Einstein obtained from his hydrodynamical method and from Planck's black-body radiation law. For Einstein, this discrepancy was particularly significant in view of what he regarded as the problematic nature of Planck's derivation of the radiation law.

In 1909 Einstein drew Perrin's attention to his hydrodynamical method for determining the size of solute molecules. He emphasized that this method allows one to take into account the volume of any water molecules attached to the solute molecules, and suggested its application to the suspensions studied by Perrin. In the following year, an experimental study of Einstein's formula for the viscosity coefficients (eq. [1] above) was performed in Perrin's laboratory by Jacques Bancelin. Bancelin studied uniform aqueous emulsions of gamboge, prepared with the help of Perrin's method of fractional centrifugation. Bancelin confirmed that the increase in viscosity does not depend on the size of the suspended particles, but only on the fraction of the total volume that they occupy. However, he found a value for the increased viscosity that differs significantly from Einstein's prediction. Bancelin sent a report of his experiments to

Einstein, apparently citing a value of 3.9 for the coefficient of φ in eq. (1), instead of the predicted value of l.

After an unsuccessful attempt to find an error in his calculations, Einstein wrote to his student and collaborator Ludwig Hopf in January 1911: "I have checked my previous calculations and arguments and found no error in them. You would be doing a great service in this matter if you would carefully recheck my investigation. Either there is an error in the work, or the volume of Perrin's suspended substance in the suspended state is greater than Perrin believes."

Hopf found an error in the derivatives of the velocity components, which occur in the equations for the pressure components in Einstein's dissertation (see pp. 53–54 below). After correction of this error, the coefficient of φ in eq. (1) becomes 2.5.

By mid-January 1911 Einstein had informed Bancelin and Perrin of Hopf's discovery of the error in his calculations. The remaining discrepancy between the corrected factor 2.5 in eq. (1) and Bancelin's experimental value of 3.9 led Einstein to suspect that there might also be an experimental error. He asked Perrin: "Wouldn't it be possible that your mastic particles, like colloids, are in a swollen state? The influence of such a swelling 3.9/2.5 would be of rather slight influence on Brownian motion, so that it might possibly have escaped you."

On 21 January 1911, Einstein submitted his correction for publication. In the *Annalen der Physik* he presented the corrected form of some of the equations in the dissertation and recalculated Avogadro's number. He obtained a value of 6.56×10^{23} per mole, a value that is close to those derived from kinetic theory and Planck's black-body radiation formula.

Bancelin continued his experiments, with results that brought experiment and theory into closer agreement. Four months later, he presented a paper on his viscosity measurements to the French Academy of Sciences, giving a value of 2.9 as the coefficient of φ in eq. (1). Bancelin also recalculated Avogadro's number by extrapolating his results for emulsions to sugar solutions, and found a value of 7.0×10^{23} per mole.

Einstein's dissertation was at first overshadowed by his more spectacular work on Brownian motion, and it required an initiative by Einstein to bring it to the attention of his fellow scientists. But the wide variety of applications of its results ultimately made the dissertation one of his most frequently cited papers.

PAPER 1

✦

A New Determination of
Molecular Dimensions

(*Ph.D. Dissertation, University of Zurich*)

THE EARLIEST determinations of the real sizes of molecules were made possible by the kinetic theory of gases, but thus far the physical phenomena observed in liquids have not helped in ascertaining molecular sizes. No doubt this is because it has not yet been possible to surmount the hurdles that impede the development of a detailed molecular-kinetic theory of liquids. It will be shown in this paper that the size of molecules of substances dissolved in an undissociated dilute solution can be determined from the internal viscosity of the solution and of the pure solvent, and from the diffusion rate of the solute within the solvent provided that the volume of a solute molecule is large compared to the volume of a solvent molecule. This is possible because, with respect to its mobility within the solvent and its effect on the viscosity of the latter, such a molecule will behave approximately like a solid body suspended in a solvent. Thus, in the immediate vicinity of a molecule, one can apply the equations of hydrodynamics to the motion of the solvent in which the liquid is treated as homogeneous and hence its molecular structure need not be taken into consideration. We will choose a sphere as the solid body that shall represent the solute molecules.

1. How a Very Small Sphere Suspended in a Liquid Influences Its Motion

Let us base our discussion on an incompressible homogeneous liquid with a coefficient of viscosity k, whose velocity components u, v, w are given as functions of the coordinates x, y, z and of time. At an arbitrary point x_0, y_0, z_0, let us think of the functions u, v, w as functions of $x - x_0$, $y - y_0$, $z - z_0$ expanded in a Taylor's series, and of a region G around this point so small that within it only the linear terms of this expansion need be considered. As is well known, the motion of the liquid within G can then be regarded as a superposition of three motions:

1. A parallel displacement of all particles of the liquid without a change in their relative positions;
2. A rotation of the liquid without a change in the relative positions of the particles of the liquid;
3. A dilational motion in three mutually perpendicular directions (the principal axes of dilation).

Let us now assume that in region G there is a spherical rigid body whose center lies at the point x_0, y_0, z_0 and whose dimensions are very small compared with those of region G. We further assume that the motion is so slow that the kinetic energy of the sphere as well as that of the liquid can be neglected. We also assume that the velocity components of a surface element of the sphere coincide with the corresponding velocity components of the adjacent liquid particles, i.e., that the contact layer (imagined to be continuous) also displays a coefficient of viscosity that is not infinitesimally small.

It is obvious that the sphere simply takes part in the partial motions 1 and 2 without altering the motion of neighboring particles, since the liquid moves like a rigid body in these partial motions and since we have neglected the effects of inertia.

However, motion 3 does get altered by the presence of the sphere, and our next task will be to investigate the effect of the sphere on this motion of the liquid. If we refer motion 3 to a coordinate system whose axes are parallel to the principal axes of dilation and set

$$x - x_0 = \xi,$$
$$y - y_0 = \eta,$$
$$z - z_0 = \zeta,$$

we can describe the above motion, if the sphere is not present, by the equations

$$\begin{cases} u_0 = A\xi, \\ v_0 = B\eta, \\ w_0 = C\zeta; \end{cases} \tag{1}$$

A, B, C are constants that, because the liquid is incompressible, satisfy the condition

$$A + B + C = 0. \tag{2}$$

If, now, a rigid sphere of radius P is introduced at the point x_0, y_0, z_0, the motion of the liquid around it will change. We will, for convenience, call P "finite," but all the values of

ξ, η, ζ, for which the liquid motion is no longer noticeably altered by the sphere, we will call "infinitely large."

Because of the symmetry of the motion of the liquid being considered, it is clear that during this motion the sphere can perform neither a translation nor a rotation, and we obtain the boundary conditions

$$u = v = w = 0 \text{ when } \rho = P,$$

where

$$\rho = \sqrt{\xi^2 + \eta^2 + \zeta^2} > 0.$$

Here u, v, w denote the velocity components of this motion (changed by the sphere). If we set

$$\begin{aligned}
u &= A\xi + u_1, \\
v &= B\eta + v_1, \\
w &= C\zeta + w_1,
\end{aligned} \tag{3}$$

the velocities u_1, v_1, w_1 would have to vanish at infinity, since at infinity the motion represented in equations (3) should reduce to that represented by equations (1).

The functions u, v, w have to satisfy the equations of hydrodynamics, including viscosity and neglecting inertia. Thus the following equations will hold:[1]

$$\begin{cases}
\dfrac{\delta p}{\delta \xi} = k\Delta u \dfrac{\delta p}{\delta \eta} = k\Delta v \dfrac{\delta p}{\delta \zeta} = \Delta w,^{[1]} \\[2mm]
\dfrac{\delta u}{\delta \xi} + \dfrac{\delta v}{\delta \eta} + \dfrac{\delta w}{\delta \zeta} = 0,
\end{cases} \tag{4}$$

[1] G. Kirchhoff, *Vorlesungen über Mechanik*, 26. Vorl. (*Lectures on Mechanics*, Lecture 26).

DETERMINATION OF MOLECULAR DIMENSIONS

where Δ denotes the operator

$$\frac{\delta^2}{\delta\xi^2} + \frac{\delta^2}{\delta\eta^2} + \frac{\delta^2}{\delta\zeta^2}$$

and p the hydrostatic pressure.

Since equations (1) are solutions of equations (4) and the latter are linear, according to (3) the quantities u_1, v_1, w_1 must also satisfy equations (4). I determined u_1, v_1, w_1, and p by a method given in section 4 of the Kirchhoff lectures mentioned above[2] and found

$$
\begin{cases}
p = -\tfrac{5}{3}kP^3 \left\{ A\dfrac{\delta^2\left[\tfrac{1}{\rho}\right]}{\delta\xi^2} + B\dfrac{\delta^2\left[\tfrac{1}{\rho}\right]}{\delta\eta^2} + C\dfrac{\delta^2\left[\tfrac{1}{\delta}\right]}{\delta\zeta^2} \right\} + \text{const.}, \\[3mm]
u = A\xi - \dfrac{5}{3}P^3 A\dfrac{\xi}{\rho^3} - \dfrac{\delta D}{\delta\xi}, \\[3mm]
v = B\eta - \dfrac{5}{3}P^3 B\dfrac{\eta}{\rho^3} - \dfrac{\delta D}{\delta\eta}, \\[3mm]
w = C\zeta - \dfrac{5}{3}P^3 C\dfrac{\zeta}{\rho^3} - \dfrac{\delta D}{\delta\zeta},
\end{cases}
\qquad (5)^{[5]}
$$

[2] "From equations (4) it follows that $\Delta p = 0$. If we take p in accordance with this condition and determine a function V that satisfies the equation

$$\Delta V = \tfrac{1}{k}p,$$

then equations (4) are satisfied if one sets

$$u = \frac{\delta V}{\delta\xi} + u', \quad v = \frac{\delta V}{\delta\eta} + v', \quad w = \frac{\delta V}{\delta\zeta} + w',$$

and chooses u', v', w' such that $\Delta u' = 0$, $\Delta v' = 0$, $\Delta w' = 0$, and

$$\frac{\delta u'}{\delta\xi} + \frac{\delta v'}{\delta\eta} + \frac{\delta w'}{\delta\zeta} = -\frac{1}{k}p."$$

Now, if one sets

$$\frac{p}{k} = 2c\frac{\delta^2\frac{1}{p}}{\delta\xi^3}, \text{ [2]}$$

49

where

$$
\begin{cases}
D = & A\left\{ \tfrac{5}{6}\, p^3 \dfrac{\delta^2 \rho}{\delta \xi^2} + \tfrac{1}{6} P^5 \dfrac{\delta^2\left(\frac{1}{\rho}\right)}{\delta \xi^2} \right\} \\[2ex]
& + B\left\{ \tfrac{5}{6}\, p^3 \dfrac{\delta^2 \rho}{\delta \eta^2} + \tfrac{1}{6} P^5 \dfrac{\delta^2\left(\frac{1}{\rho}\right)}{\delta \eta^2} \right\} \\[2ex]
& + C\left\{ \tfrac{5}{6} p^3 \dfrac{\delta^2 \rho}{\delta \zeta^2} + \tfrac{1}{6} P^5 \dfrac{\delta^2\left(\frac{1}{\rho}\right)}{\delta \zeta^2} \right\}.
\end{cases}
\tag{5a}
$$

It can easily be proved that equations (5) are solutions of equations (4). Since

$$
\Delta \xi = 0, \quad \Delta \frac{1}{\rho} = 0, \quad \Delta \rho = \frac{2}{\rho}
$$

and

$$
\Delta\left(\frac{\xi}{\rho^3}\right) = -\frac{\delta}{\delta \xi}\left\{ \Delta\left(\frac{1}{\rho}\right) \right\} = 0,
$$

we get

$$
\begin{aligned}
k\Delta u &= -k\frac{\delta}{\delta \xi}\{\Delta D\} \\
&= -k\frac{\delta}{\delta \xi}\left\{ \tfrac{5}{3} P^3 A \dfrac{\delta^2 \frac{1}{\rho}}{\delta \xi^2} + \tfrac{5}{3} P^3 B \dfrac{\delta^2 \frac{1}{\rho}}{\delta \eta^2} + \cdots \right\}.
\end{aligned}
$$

---

and, in accordance with this,

$$
V = c\, \frac{\delta^2 \rho}{\delta \xi^3} + b\, \frac{\delta^2 \frac{1}{\rho}}{\delta \xi^2} + \frac{a}{2}\left[\xi^2 - \frac{\eta^2}{2} - \frac{\zeta^2}{2} \right]^{[3]}
$$

and

$$
u' = -2c\, \frac{\delta \frac{1}{\delta}}{\delta \xi}, \quad v' = 0, \quad w' = 0,^{[4]}
$$

then the constants a, b, c can be determined such that $u = v = w = 0$ for $\rho = P$. By superposing three such solutions, we get the solution given in equations (5) and (5a).

However, according to the first of equations (5), the last of the expressions we obtained is identical to $\frac{\delta n}{\delta \xi}$.[6] In the same way, it can be shown that the second and third of equations (4) are satisfied. Further, we get

$$\frac{\delta u}{\delta \xi} + \frac{\delta v}{\delta \eta} + \frac{\delta w}{\delta \xi} = (A + B + C)$$
$$+ \tfrac{5}{3}P^3 \left\{ A \frac{\delta^2 \left(\frac{1}{\rho} \right)}{\delta \xi^2} + B \frac{\delta^2 \left(\frac{1}{\rho} \right)}{\delta \eta^2} + C \frac{\delta^2 \left(\frac{1}{\rho} \right)}{\delta \zeta^2} \right\} - \Delta D.$$

But since according to equation (5a)

$$\Delta D = \tfrac{5}{3}AP^3 \left\{ A \frac{\delta^2 \left(\frac{1}{\rho} \right)}{\delta \xi^2} + B \frac{\delta^2 \left(\frac{1}{\rho} \right)}{\delta \eta^2} + C \frac{\delta^2 \left(\frac{1}{\rho} \right)}{\delta \zeta^2} \right\},$$

it follows that the last of equations (4) is satisfied as well. As far as the boundary conditions are concerned, at infinitely large ρ our equations for u, v, w reduce to equations (1). By inserting the value of D from equation (5a) into the second of equations (5), we get

$$u = A\xi - \tfrac{5}{2}\frac{P^3}{\rho^6}\xi(A\xi^2 + B\eta^2 + C\zeta^2)^{[7]}$$
$$+ \tfrac{5}{2}\frac{P^5}{\rho^7}\xi(A\xi^2 + B\eta^2 + C\zeta^2) - \frac{P^5}{\rho^5}A\xi. \tag{6}$$

We see that u vanishes for $\rho = P$. For reasons of symmetry, the same holds for v and w. We have now demonstrated that equations (5) satisfy equations (4) as well as the boundary conditions of the problem.

It can also be demonstrated that equations (5) are the only solution of equations (4) that is compatible with the boundary conditions of our problem. The proof will only be outlined here. Assume that in a finite region the velocity components u, v, w of a liquid satisfy equations (4). If there existed yet another solution U, V, W for equations (4) in which $U = u$, $V = v$, $W = w$ at the boundaries of the region in question, then $(U - u, V - v, W - w)$ would be a solution for equations (4) in which the velocity components vanish at the boundary. Thus no mechanical work is supplied to the liquid in the region in question. Since we have neglected the kinetic energy of the liquid, it follows that in this volume the work converted to heat is also zero. This leads to the conclusion that in the entire space we must have $u = u_1$, $v = v_1$, $w = w_1$ if the region is at least partly bounded by stationary walls.[8] By passing to the limit, this result can also be extended to a case where the region is infinite, as in the case considered above. One can thus show that the solution found above is the only solution to the problem.

We now draw a sphere of radius R around point x_0, y_0, z_0, with R infinitely large compared to P, and calculate the energy (per unit time) that is converted to heat in the liquid inside the sphere. This energy W is equal to the mechanical work done on the liquid. If X_n, Y_n, Z_n denote the components of the pressure exerted on the surface of the sphere of radius R, we have

$$W = \int (X_n u + Y_n v + Z_n w)\, ds,$$

where the integral is to be extended over the surface of the sphere of radius R. We have here

$$X_n = -\left(X\xi \frac{\xi}{\rho} + X\eta \frac{\eta}{\rho} + X\zeta \frac{\zeta}{\rho} \right),{}^{[9]}$$

$$Y_n = -\left(Y\xi \frac{\xi}{\rho} + Y\eta \frac{\eta}{\rho} + Y\zeta \frac{\zeta}{\rho} \right),$$

$$Z_n = -\left(Z\xi \frac{\xi}{\rho} + Z\eta \frac{\eta}{\rho} + Z\zeta \frac{\zeta}{\rho} \right),$$

where

$$X_\xi = p - 2k\frac{\delta u}{\delta \xi}, \qquad Y_\zeta = Z_\eta = -k\left(\frac{\delta v}{\delta \zeta} + \frac{\delta w}{\delta \eta}\right),$$

$$Y_\eta = p - 2k\frac{\delta v}{\delta \eta}, \qquad Z_\xi = X_\zeta = -k\left(\frac{\delta w}{\delta \xi} + \frac{\delta u}{\delta \zeta}\right),$$

$$Z_\zeta = p - 2k\frac{\delta w}{\delta \zeta}, \qquad X_\eta = Y_\xi = -k\left(\frac{\delta u}{\delta \eta} + \frac{\delta v}{\delta \xi}\right).$$

The expressions for u, v, w become simpler if we take into account that for $\rho = R$ the terms with the factor P^5/ρ^5 vanish in comparison to those with the factor P^3/ρ^3. We have to set

$$\begin{cases} u = A\xi - \frac{5}{2}P^3\dfrac{\xi(A\xi^2 + B\eta^2 + C\zeta^2)}{\rho^5}, \\[2mm] v = B\eta - \frac{5}{2}P^3\dfrac{\eta(A\xi^2 + B\eta^2 + C\zeta^2)}{\rho^5}, \\[2mm] w = C\zeta - \frac{5}{2}P^3\dfrac{\zeta(A\xi^2 + B\eta^2 + C\zeta^2)}{\rho^5}. \end{cases} \qquad \text{(6a)[10]}$$

For p we obtain from the first of equations (5), by similar neglect of terms,

$$p = -5kP^3\frac{A\xi^2 + B\eta^2 + C\zeta^2}{\rho^5} + \text{const.[11]}$$

Now we obtain

$$X_\xi = -2kA + 10kP^3\frac{A\xi^2}{\rho^5} - 25kP^3\frac{\xi^2(A\xi^2 + B\eta^2 + C\zeta^2)}{\rho^7}^{[12]}$$

$$X_\eta = +10kP^3\frac{A\xi\eta}{\rho^5} - 25kP^3\frac{\eta^2(A\xi^2 + B\eta^2 + C\zeta^2)}{\rho^7}^{[13]}$$

$$X_\zeta = +10kP^3\frac{A\xi\zeta}{\rho^5} + 25kP^3\frac{\zeta^2(A\xi^2 + B\eta^2 + C\zeta^2)}{\rho^7},$$

and from this,

$$X_n = 2Ak\frac{\xi}{\rho} - 10AkP^3\frac{\xi}{\rho^4} + 25kP^3\frac{\xi(A\xi^2 + B\eta^2 + C\zeta^2)}{\rho^6}. \quad [14]$$

With the help of the expressions for Y_n and Z_n derived by cyclic permutation, and ignoring all terms that contain the ratio P/ρ in higher than the third power, we get[15]

$$X_n u + Y_n v + Z_n w + \frac{2k}{\rho}(A^2\xi^2 + B^2\eta^2 + C^2\zeta^2)$$
$$-10k\frac{P^3}{\rho^4}(A^2\xi^2 + . + .) + 20k\frac{P^3}{\rho^6}(A\xi^2 + . + .)^2.$$

If we integrate over the sphere and take into account that

$$\int ds = 4R^2\pi,$$

$$\int \xi^2 \, ds = \int \eta^2 \, ds = \int \zeta^2 \, ds = \tfrac{4}{3}\pi R^4,$$

$$\int \xi^4 \, ds = \int \eta^4 \, ds = \int \zeta^4 \, ds = \tfrac{4}{5}\pi R^6,$$

$$\int \eta^2\zeta^2 \, ds = \int \zeta^2\xi^2 \, ds = \int \xi^2\eta^2 \, ds = \tfrac{4}{15}\pi R^6, \quad [16]$$

$$\int (A\xi^2 + B\eta^2 + C\zeta^2)^2 \, ds = \tfrac{4}{15}\pi R^6(A^2 + B^2 + C^2), \quad [17]$$

we get[18]

$$W = \tfrac{8}{3}\pi R^3 k\delta^2 - \tfrac{8}{3}\pi P^3 k\delta^2 = 2\delta^2 k(V - \Phi), \quad (7)$$

where we set

$$\delta = A^2 + B^2 + C^2, \quad [19]$$
$$\tfrac{4}{3}\pi R^3 = V$$

and

$$\tfrac{4}{3}\pi P^3 = \Phi.$$

If the suspended sphere were not present ($\Phi = 0$), we would obtain

$$W_0 = 2\delta^2 kV \qquad (7a)$$

for the energy dissipated in volume V. Thus, the presence of the sphere decreases the energy dissipated by $2\delta^2 k\Phi$. It is noteworthy that the effect of the suspended sphere on the quantity of energy dissipated is exactly the same as it would be if the presence of the sphere did not affect the motion of the liquid around it at all.[20]

2. Calculation of the Coefficient of Viscosity of a Liquid in Which Very Many Irregularly Distributed Small Spheres Are Suspended

In the previous section we considered the case where, in a region G of the order of magnitude defined earlier, a sphere is suspended that is very small compared with the region, and we investigated how this sphere affects the motion of the liquid. We are now going to assume that region G contains innumerably many randomly distributed spheres of equal radius, and that this radius is so small that the combined volume of all of the spheres is very small compared to the region G. Let the number of spheres per unit volume be n, where, up to negligibly small terms, n is constant throughout the liquid.

Again, we begin with the motion of a homogeneous liquid without any suspended spheres and consider again the most general dilational motion. If no spheres are present, an appropriate choice of the coordinate system will permit us to represent the velocity components u_0, v_0, w_0 at an arbitrary point x, y, z of G by the equations

$$u_0 = Ax,$$
$$v_0 = By,$$
$$w_0 = Cz,$$

where

$$A + B + C = 0.$$

A sphere suspended at point x_ν, y_ν, z_ν will affect this motion in a way that is evident from equation (6).[21] Since we are choosing the average distance between neighboring spheres to be large compared to their radius, and consequently the additional velocity components arising from all the suspended spheres are very small compared to u_0, v_0, w_0, we obtain for the velocity components u, v, w in the liquid, after taking into account the suspended spheres and neglecting terms of higher orders,

$$\left\{ \begin{aligned} u &= Ax - \sum \left\{ \begin{aligned} & \frac{5}{2} \frac{P^3}{\rho_\nu^2} \frac{\xi_\nu(A\xi_\nu^2 + B\eta_\nu^2 + C\zeta_\nu^2)}{\rho_\nu^3} \\ & - \frac{5}{2} \frac{P^5}{\rho_\nu^4} \frac{\xi_\nu(A\xi_\nu^2 + B\eta_\nu^2 + C\zeta_\nu^2)}{\rho_\nu^3} + \frac{P^5}{\rho_\nu^4} \frac{A\xi_\nu}{\rho_\nu} \end{aligned} \right\}, \\ v &= By - \sum \left\{ \begin{aligned} & \frac{5}{2} \frac{P^3}{\rho_\nu^2} \frac{\eta_\nu(A\xi_\nu^2 + B\eta_\nu^2 + C\zeta_\nu^2)}{\rho_\nu^3} \\ & - \frac{5}{2} \frac{P^5}{\rho_\nu^4} \frac{\eta_\nu(A\xi_\nu^2 + B\eta_\nu^2 + C\zeta_\nu^2)}{\rho_\nu^3} + \frac{P^5}{\rho_\nu^4} \frac{B\eta_\nu}{\rho_\nu} \end{aligned} \right\}, \\ w &= Cz - \sum \left\{ \begin{aligned} & \frac{5}{2} \frac{P^3}{\rho_\nu^2} \frac{\zeta_v(A\xi_v^2 + B\eta_\nu^2 + C\zeta_\nu^2)}{\rho_\nu^3} \\ & - \frac{5}{2} \frac{P^5}{\rho_\nu^4} \frac{\zeta_\nu(A\xi_\nu^2 + B\eta_\nu^2 + C\zeta_\nu^2)}{\rho_\nu^3} + \frac{P^5}{\rho_\nu^4} \frac{C\zeta_\nu}{\rho_\nu} \end{aligned} \right\}, \end{aligned} \right. \quad (8)$$

where the sum is to be extended over all spheres in the region G and where we have set

$$\xi_\nu = x - x_\nu,$$
$$\eta_\nu = y - y_\nu, \qquad \rho_\nu = \sqrt{\xi_\nu^2 + \eta_\nu^2 + \zeta_\nu^2}.$$
$$\zeta_\nu = z - z_\nu,$$

x_ν, y_ν, z_ν are the coordinates of the centers of the spheres. Furthermore, from equations (7) and (7a) we conclude that, up to infinitesimally small quantities of higher order, the presence of each sphere results in a decrease of heat production by $2\delta^2 k\Phi$ per unit time[22] and that the energy converted to heat in region G has the value

$$W = 2\delta^2 k - 2n\delta^2 k\Phi$$

per unit volume, or

$$W = 2\delta^2 k(1 - \varphi), \tag{7b}$$

where φ denotes the fraction of the volume that is occupied by the spheres.

Equation (7b) gives the impression that the coefficient of viscosity of the inhomogeneous mixture of liquid and suspended spheres (in the following called "mixture" for short) is smaller than the coefficient of viscosity k of the liquid.[23] However, this is not so, since A, B, C are not the values of the principal dilations of the liquid flow represented by equations (8); we will call the principal dilations of the mixture A^*, B^*, C^*. For reasons of symmetry, it follows that the directions of the principal dilations of the mixture are parallel to the directions of the principal dilations A, B, C, i.e., to the coordinate axes. If we write equations (8) in the form

$$u = Ax + \sum u_\nu,$$
$$v = By + \sum v_\nu,$$
$$z = Cz + \sum w_\nu,$$

we get

$$A^* = \left(\frac{\delta u}{\delta x}\right)_{x=0} = A + \sum\left(\frac{\delta u_v}{\delta x}\right)_{x=0} = A - \sum\left(\frac{\delta u_v}{\delta x_v}\right)_{x=0}.$$

If we exclude the immediate surroundings of the individual spheres, we can omit the second and third terms in the expressions for u, v, w and thus obtain for $x = y = z = 0$:

$$\begin{cases} u_\nu = -\frac{5}{2}\frac{P^3}{r_\nu^2}\frac{x_\nu(Ax_\nu^2 + By_\nu^2 + Cz_\nu^2)}{r_\nu^3}, \\[2ex] v_\nu = -\frac{5}{2}\frac{P^3}{r_\nu^2}\frac{y_\nu(Ax_\nu^2 + By_\nu^2 + Cz_\nu^2)}{r_\nu^3}, \\[2ex] w_\nu = -\frac{5}{2}\frac{P^3}{r_\nu^2}\frac{x(Ax_\nu^2 + By_\nu^2 + Cz_\nu^2)}{r_\nu^3}, \end{cases} \qquad (9)^{[24]}$$

where we have set

$$r_\nu = \sqrt{x_\nu^1 + y_\nu^2 + z_\nu^2} > 0.$$

We extend the summation over the volume of a sphere K of very large radius R whose center lies at the coordinate origin. Further, if we consider the *irregularly* distributed spheres as being *uniformly* distributed and replace the sum with an integral, we obtain[25]

$$A^* = A - n\int_K \frac{\delta u_\nu}{\delta x_\nu}\, dx_\nu\, dy_\nu\, dz_\nu,$$
$$= A - n\int \frac{u_\nu x_\nu}{r_\nu}\, ds,$$

where the last integral extends over the surface of the sphere K. Taking into account (9), we find that

$$A^* = A - \frac{5}{2}\frac{P^3}{R^6}n \int x_0^2(Ax_0^2 + By_0^2 + Cz_0^2)\,ds,$$

$$= A - n\left(\frac{4}{3}P^3\pi\right)A = A(1 - \varphi).$$

Analogously,

$$B^* = B(1 - \varphi),$$
$$C^* = C(1 - \varphi).$$

If we set

$$\delta^{*2} = A^{*2} + B^{*2} + C^{*2},\text{[26]}$$

then, neglecting infinitesimally small terms of higher order,

$$\delta^{*2} = \delta^2(1 - 2\varphi).$$

For the heat developed per unit time and volume, we found[27]

$$W^* = 2\delta^2 k(1 - \varphi).$$

If k^* denotes the coefficient of viscosity of the mixture, we have

$$W^* = 2\delta^{*2}k^*.$$

The last three equations yield, neglecting infinitesimal quantities of higher order,

$$k^* = k(1 + \varphi).\text{[28]}$$

Thus we obtain the following result:

If very small rigid spheres are suspended in a liquid, the coefficient of viscosity increases by a fraction that is equal to the total volume of the spheres suspended in a unit volume, provided that this total volume is very small.[29]

3. On the Volume of a Dissolved Substance Whose Molecular Volume Is Large Compared to That of the Solvent

Consider a dilute solution of a substance that does not dissociate in the solution. Let a molecule of the dissolved substance be large compared to a molecule of the solvent and be considered as a rigid sphere of radius P. We can then apply the result obtained in section 2. If k^* denotes the coefficient of viscosity of the solution and k that of the pure solvent, we have

$$\frac{k^*}{k} = 1 + \varphi,$$

where φ is the total volume of the molecules per unit volume of the solution.[30]

We wish to calculate φ for a 1% aqueous solution of sugar. According to Burkhard's observations (Landolt and Börnstein's *Tables*), $k^*/k = 1.0245$ (at 20°C) for a 1% aqueous sugar solution, hence $\varphi = 0.0245$ for (almost exactly) 0.01 g of sugar. Thus, one gram of sugar dissolved in water has the same effect on the coefficient of viscosity as do small suspended rigid spheres of a total volume of 2.45 cm^3.[31] This consideration neglects the effect exerted on the viscosity of the solvent by the osmotic pressure resulting from the dissolved sugar.

Let us remember that 1 g of solid sugar has a volume of 0.61 cm^3. This same volume is also found for the specific volume s of sugar in solution if one considers the sugar solution as a *mixture* of water and sugar in dissolved form. I.e., the density of a 1% aqueous sugar solution (relative to water of the same temperature) at 17.5° is 1.00388. Hence we have (neglecting the difference between the density of water at $4°$ and at $17.5°$)

$$\frac{1}{1.00388} = 0.99 + 0.01\,s,$$

and thus

$$s = 0.61.$$

Thus, while the sugar solution behaves like a mixture of water and solid sugar with respect to its density, the effect on viscosity is four times larger than what would result from the suspension of the same amount of sugar.[32] It seems to me that, from the point of view of molecular theory, this result can only be interpreted by assuming that a sugar molecule in solution impedes the mobility of the water in its immediate vicinity, so that an amount of water whose volume is about three times larger than the volume of the sugar molecule is attached to the sugar molecule.[33]

Hence we may say that a dissolved molecule of sugar (i.e., the molecule together with the water attached to it) behaves hydrodynamically like a sphere with a volume of $2.45 \cdot 342/N$ cm^3, where 342 is the molecular weight of sugar and N is the number of actual molecules in one gram-molecule.[34]

4. ON THE DIFFUSION OF AN UNDISSOCIATED SUBSTANCE IN A LIQUID SOLUTION

Let us consider a solution of the kind discussed in section 3. If a force K acts upon a molecule, which we assume to be a sphere with radius P, the molecule will move with a velocity ω, which is determined by P and the coefficient of viscosity

k of the solvent. Indeed, the following equation holds:[3]

$$\omega = \frac{K}{6\pi k P}. \tag{1}$$

We use this relation to calculate the coefficient of diffusion of an undissociated solution. If p is the osmotic pressure of the dissolved substance, the only motion-producing force in such a dilute solution, then the force acting on the dissolved substance per unit volume of solution in the direction of the X-axis equals $-\delta p/\delta x$. If there are ρ grams per unit volume, and m is the molecular weight of the dissolved substance and N the number of actual molecules in one gram-molecule, then $(\rho/m) \cdot N$ is the number of (actual) molecules per unit volume, and the force exerted on a molecule by virtue of the concentration gradient is

$$K = -\frac{m}{\rho N}\frac{\delta p}{\delta x}. \tag{2}$$

If the solution is sufficiently dilute, the osmotic pressure is given by the equation:

$$p = \frac{R}{m}\rho T, \tag{3}$$

where T is the absolute temperature and $R = 8.31 \cdot 10^7$. From equations (1), (2), and (3) we obtain for the migration velocity of the dissolved substance

$$\omega = -\frac{RT}{6\pi k}\frac{1}{NP}\frac{1}{\rho}\frac{\delta \rho}{\delta x}. \tag{4}$$

[3] G. Kirchhoff, *Vorlesungen über Mechanik*, 26. Vorl. (*Lectures on Mechanics*, Lecture 26), equation (22).

Finally, the amount of the substance passing per unit time through a unit cross section in the direction of the X-axis is

$$\omega\rho = -\frac{RT}{6\pi k} \cdot \frac{1}{NP}\frac{\delta\rho}{\delta x}.$$

Hence, we obtain for the coefficient of diffusion D

$$D = \frac{RT}{6nk} \cdot \frac{1}{NP}. \text{[35]}$$

Thus, from the coefficients of diffusion and viscosity of the solvent we can calculate the product of the number N of actual molecules in one gram-molecule and the hydrodynamically effective molecular radius P.

In this derivation the osmotic pressure has been treated as a force acting on the individual molecules, which obviously does not agree with the viewpoint of the kinetic molecular theory; since in our case—according to the latter—the osmotic pressu.e must be conceived as only an apparent force. However, this difficulty disappears when one considers that the (apparent) osmotic forces that correspond to the concentration gradients in the solution may be kept in (dynamic) equilibrium by means of numerically equal forces acting on the individual molecules in the opposite direction, which can easily be seen by thermodynamic methods.

The osmotic force acting on a unit mass $-\frac{1}{\rho}\frac{\delta p}{\delta x}$ can be counterbalanced by the force $-P_x$ (exerted on the individual dissolved molecules) if

$$-\frac{1}{\rho}\frac{\delta p}{\delta x} - P_x = 0.$$

Thus, if one imagines that (per unit mass) the dissolved substance is acted upon by two sets of forces P_x and $-P_x$ that mutually cancel out each other, then $-P_x$ counterbalances the osmotic pressure, leaving only the force P_x, which is numerically equal to the osmotic pressure, as the cause of motion. The difficulty mentioned above has thus been eliminated.[4]

5. Determination of Molecular Dimensions with the Help of the Obtained Relations

We found in section 3 that

$$\frac{k^*}{k} = 1 + \varphi = 1 + n \cdot \tfrac{4}{3}\pi P^3, {}^{[36]}$$

where n is the number of dissolved molecules per unit volume and P is the hydrodynamically effective radius of the molecule. If we take into account that

$$\frac{n}{N} = \frac{\rho}{m},$$

where ρ denotes the mass of the dissolved substance per unit volume and m its molecular weight, we get

$$NP^3 = \frac{3}{4\pi}\frac{m}{\rho}\left(\frac{k^*}{k} - 1\right).{}^{[37]}$$

On the other hand, we found in section 4 that

$$NP = \frac{RT}{6\pi k}\frac{1}{D}.$$

These two equations enable us to calculate separately the quantities P and N, of which N must be independent of the nature of the solvent, the dissolved substance, and the temperature, if our theory agrees with the facts.

---

[4]A detailed presentation of this line of reasoning can be found in *Ann. d. Phys.* 17 (1905): 549. [See also this volume, paper 2, p. 86.]

We will carry out the calculation for an aqueous solution of sugar. From the data on the viscosity of the sugar solution cited earlier, it follows that at 20°C,

$$NP^3 = 200.^{[38]}$$

According to the experiments of Graham (as calculated by Stefan), the diffusion coefficient of sugar in water is 0.384 at 9.5°C, if the day is chosen as the unit of time. The viscosity of water at 9.5° is 0.0135. We will insert these data in our formula for the diffusion coefficient, even though they have been obtained using 10% solutions, and strict validity of our formula cannot be expected at such high concentrations. We obtain

$$NP = 2.08 \cdot 10^{16}.$$

Neglecting the differences between the values of P at 9.5° and 20°, the values found for NP^3 and NP yield

$$P = 9.9 \cdot 10^{-8} \text{ cm},$$
$$N = 2.1 \cdot 10^{23}.$$

The value found for N shows satisfactory agreement, in order of magnitude, with values found for this quantity by other methods.[39]

(Bern, 30 April 1905)

EDITORIAL NOTES

[1]A factor k is missing on the right-hand side of the last equation in this line; this error is corrected in Albert Einstein, "Eine neue Bestimmung der Moleküldimensionen," *Ann. d. Phys.* 19 (1906), pp. 289–305, cited hereafter as *Einstein 1906*. Note that $\frac{\delta}{\delta}$ denotes partial differentiation (modern $\frac{\partial}{\partial}$).

[2]The denominator on the right-hand side should be $\delta\xi^2$; this error is corrected in ibid.

[3]The denominator of the first term on the right-hand side should be $\delta\xi^2$; this error is corrected in ibid. A reprint of this article in the Einstein Archive shows marginalia and interlineations in Einstein's hand, the first of which refer to this and the following equation. The term "$+g\frac{1}{p}$" was added to the right-hand side of the equations for V and then canceled. These marginalia and interlineations are presumably part of Einstein's unsuccessful attempt to find a calculational error; see note 13 below.

[4]The equation for u' should be, as corrected in ibid., $u' = -2c\frac{\delta\frac{1}{p}}{\delta\xi}$. In the reprint mentioned in note 3, the first derivative with respect to ξ was changed to a second derivative and then changed back to a first derivative. At the bottom of the page, the following equations are written:

$$b = -1/12\ P^5 a$$
$$c = -5/12\ P^3 a$$
$$g = 2/3\ P^3 a.$$

[5]The numerator of the last term in the curly parentheses should be "$\delta^2(1/\rho)$," as corrected in ibid.

[6]$\frac{\delta n}{\delta\xi}$ should be $\frac{\delta p}{\delta\xi}$, as corrected in Einstein, *Untersuchungen über die Theorie der 'Brownschen Bewegung'* (ed. Reinhold Fürth. Ostwald's Klassiker der exakten Wissenschaften, no. 199. Leipzig: Akademische Verlagsgesellschaft, 1922); cited hereafter as *Einstein 1922*.

[7]The factor preceding the first parenthesis should be, as corrected in *Einstein 1906*,

$$-5/2\frac{P^3}{\rho^5}.$$

[8]The equations should be $u = U$, $v = V$, $w = W$.

[9]$X\xi$, $X\eta$, $X\zeta$ should be X_ξ, etc., as corrected in *Einstein 1906*.

[10]In Einstein's reprint (see note 3), the term $+\frac{5}{6}P^3\frac{A\xi}{\rho^3}$ is added to the right-hand side of the first equation. After the last terms of the second and third equations, series of dots are added. These interlineations are presumably related to the marginal calculations indicated in note 3.

[11]In Einstein's reprint (see note 3), the term $+5\,k\,P^3\frac{1}{\rho^3}$ is added to the right-hand side of this equation. This interlineation is presumably related to the marginal calculations referred to in note 3.

[12]In Einstein's reprint (see note 3), the term $-\frac{5}{3} k P^3 A\left(\frac{1}{\rho^3} - 9\frac{\xi}{\rho^5}\right)$ is added to the right-hand side of this equation. This addition is presumably related to the marginal calculations referred to in note 3.

[13]This equation and the subsequent one are incorrect. Apart from minor errors, they contain a calculational error bearing on the numerical factors. In *Einstein 1906*, +25 in front of the last term in the equation for X_ζ is changed to −25. In Einstein's reprint (see note 3), the factor ζ^2 in the last term on the right-hand side of this equation is corrected to $\xi\zeta$, and the factor η^2 in front of the parenthesis in the last term on the right-hand side of the equation for X_η is corrected to $\xi\eta$. The calculational error that is also contained in these equations, and some of its consequences, are corrected in "Berichtigung zu meiner Arbeit: 'Eine Neue Bestimmung der Moleküldimensionen,'" *Collected Papers*, vol. 3, doc. 14, pp. 416–417. The corrections are integrated into the text of the reprint of this paper in *Einstein 1922*. The correct equations are:

$$X_\eta = +5kP^3 \, \frac{(A+B)\xi\eta}{\rho^5}$$
$$-25kP^3 \frac{\xi\eta(A\xi^2 + B\eta^2 + C\zeta^2)}{\rho^7}$$
$$X_\zeta = +5kP^3 \frac{(A+C)\xi\zeta}{\rho^5}$$
$$-25kP^3 \frac{\xi\zeta(A\xi^2 + B\eta^2 + C\zeta^2)}{\rho^7}.$$

[14]−10 should be replaced by −5 and 25 by 20 (see previous note).

[15]The third + sign should be replaced by = as corrected in *Einstein 1922*. −10 should be replaced by −5, and 20 by 15 (see note 13).

[16]In Einstein's reprint (see note 3), the factor 4/15 was changed to 8/15 and then changed back to 4/15.

[17]4/15 should be replaced by 8/15 as corrected in Einstein's reprint (see note 3).

[18]This equation should be (see note 13):

$$W = 8/3\pi R^3 k\delta^2 + 4/3\pi P^3 k\delta^2$$
$$= 2\delta^2 k(V + \Phi/2).$$

[19]δ should be δ^2. This correction is made in Einstein's reprint (see note 3).

[20]It follows from the correction to eq. (7) that the dissipated energy is actually increased by half this amount. The statement in the text is only partially corrected in *Einstein 1922*; the amount is correctly given but still

described as a diminution. The final sentence of this paragraph, which no longer applies to the corrected calculation, is omitted from *Einstein 1922*.

[21]The point should be denoted by x_ν, y_ν, z_ν, as corrected in *Einstein 1906*.

[22]The heat production per unit time is actually increased by $\delta^2 k\Phi$. The correct equations are thus (see note 13): $W = 2\delta^2 k + n\delta^2 k\Phi$, and $W = 2\delta^2 k(1 + \phi/2)$.

[23]The following two sentences are revised in *Einstein 1922*: "In order to calculate from equation (7b) the coefficient of friction of an inhomogeneous mixture of fluid and suspended spheres (in the following called 'mixture' for short) that we are examining, we must further take into consideration that A, B, C are not values of the principal dilations of the motion of fluid represented in equation (8); we want to designate the principal dilations of the mixture as A^*, B^*, C^*."

[24]In this and the following two equations, the sign after $=$ should be $+$; the third equation should have z_ν instead of x in the numerator; the latter correction is made in *Einstein 1906*.

[25]The factor in front of the second term in the first equation is 5/2 (see ibid.). In deriving the second equation, Einstein used the equations in the middle of p. 54 and the fact that $A + B + C = 0$.

[26]In Einstein's reprint (see note 3), $= A^2 + B^2 + \delta^2(1 - 2\varphi)$ is added to the right-hand side of this equation and then crossed out.

[27]The correct equation is (see note 13): $W^* = 2\delta^2 k(1 + \varphi/2)$.

[28]The correct equation is (see note 13): $k^* = k(1 + 2.5\varphi)$.

[29]The fraction is actually 2.5 times the total volume of the suspended spheres (see note 13).

[30]The correct equation is (see note 13): $k^*/k = 1 + 2.5\varphi$.

[31]The correct value is 0.98 cm^3 (see note 13). The following sentence is omitted in *Einstein 1906*.

[32]The viscosity is actually one and one-half times greater (see note 13).

[33]The quantity of water bound to a sugar molecule has a volume that is actually one-half that of the sugar molecule (see note 13). The existence of molecular aggregates in combination with water was debated at that time.

[34]The volume of the sphere is actually $0.98 \cdot 342/N$ cm^3 (see note 13).

[35]The first denominator should be $6\pi k$, as corrected in *Einstein 1906*. This equation was obtained independently by William Sutherland in 1905 by a similar argument. The idea to use this formula for a determination of molecular dimensions may have occurred to Einstein as early as 1903.

[36]The correct equation is (see note 13): $k^*/k = 1 + 2.5\varphi = 1 + 2.5n\,4/3\,\pi P^3$.

[37]The correct equation has an additional factor 2/5 on the right-hand side (see note 13).

[38]For the experimental data, see p. 60. The correct value is 80 (see note 13).

[39]The values obtained by using the correct equations (see *Einstein 1922*) are $P = 6.2\ 10^{-8}$ cm; and $N = 3.3\ 10^{23}$ (per mole).

Part Two

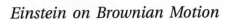

Einstein on Brownian Motion

Einstein as a student at the ETH, or shortly thereafter.
(Courtesy of Hebrew University of Jerusalem)

Einstein's study of Brownian motion constitutes one of the high points in the long tradition of research on the kinetic theory of heat and of his own contributions to this field. Some of the consequences of his work were of great significance for the development of physics in the twentieth century. Einstein's derivation of the laws governing Brownian motion, and their subsequent experimental verification by Perrin and others, contributed significantly to the acknowledgment of the physical reality of atoms by the then still-numerous skeptics. His papers on Brownian motion helped to establish the study of fluctuation phenomena as a new branch of physics. The methods he created in the course of his research prepared the way for statistical thermodynamics, later developed by Szilard and others, and for a general theory of stochastic processes.

Since at least the middle of the nineteenth century, a growing number of physicists and chemists had accepted the atomic hypothesis. The assumption that matter consists of atoms and molecules suggested a number of relations between phenomena, both physical and chemical, that are unexpected from a purely macroscopic point of view. Various methods for the determination of molecular dimensions gave values that were often in surprisingly good agreement. The physical reality of atoms was not, however, universally accepted by the end of the century. There were still some fervent opponents of the atomic hypothesis, such as Wilhelm Ostwald and Georg Helm, who called themselves "energeticists" to indicate that they regarded the concept of energy as the most fundamental ontological concept of science. Others, such as Ernst Mach, while adopting a hostile position with regard to the existence of entities not directly accessible to sense experience—in particular, atoms—admitted that atomism may have a heuristic or didactic utility. It was not uncommon, even among scientists who made explicit use of atomic assumptions in their work, to regard atomism as a mere working hypothesis.

Although at the turn of the century the atomic hypothesis was proving its heuristic value in such new areas of research as the electron theory of metals and stereochemistry, some physicists had come to regard the theory of heat as an area in which the atomic hypothesis was no longer fruitful. Einstein probably became aware of the controversy over the molecular theory of heat during his student years, when he read works by Mach, Ostwald, and Boltzmann. In 1900 Einstein finished reading Boltzmann's *Gastheorie* (1896, 1898), in which Boltzmann, presumably reacting to a dispute with Ostwald and Helm, suggested that he was isolated in his support of the kinetic theory. Although he criticized Boltzmann for a lack of emphasis on the comparison of his theory with observation, Einstein was firmly convinced of the principles of Boltzmann's theory.

In his first published attempts at independent research, Einstein took for granted the atomistic constitution of matter and of electricity. He developed a theory of molecular forces, on the basis of which he established a number of relations among observable phenomena. Einstein's interest soon shifted from the details of molecular forces to the quest for facts, "which would guarantee as much as possible the existence of atoms of definite finite size," as he later characterized this phase of his work.[1]

THE IRREGULAR movement of microscopic particles suspended in a liquid had been noted long before the botanist

Robert Brown published his careful observations in 1828, but Brown was the first to emphasize its ubiquity and to exclude its explanation as a vital phenomenon. Advances in observational technique and in theory served to eliminate a number of unsatisfactory explanations of Brownian motion by the end of the nineteenth century, if not to verify the correct one. Explanations of Brownian motion proposed after the exclusion of vital forces involved capillarity, convection currents, evaporation, interaction with light, and electrical forces. During the 1870s, the kinetic theory of heat was proposed as an explanation by several authors. A powerful argument against this explanation was developed by the cytologist Karl von Nägeli in 1879. He first used the equipartition theorem to calculate the average velocity of the molecules of the liquid, and then used the laws of elastic collision to obtain the velocity of a suspended particle. He concluded that the velocity of such a particle, because of its comparatively large mass, would be vanishingly small. William Ramsay and Louis-Georges Gouy independently tried to defend the molecular explanation of Brownian motion by assuming the existence of collective motions of large numbers of atoms in liquids, an assumption suited to the refutation of arguments such as Nägeli's.

In 1900 an entirely different way of applying the kinetic theory of heat to Brownian motion was investigated by Felix Exner, who assumed an equipartition of energy between the molecules of the liquid and the suspended particles. He calculated the velocity of the molecules on the basis of observations that he interpreted as giving the mean velocities of the suspended particles, obtaining results that were not in agreement with contemporary estimates of molecular velocities. In Exner's work there is no fundamental difference between a solute molecule and a suspended particle. Einstein arrived at a similar conclusion, but instead of emphasizing the equipartition theorem, he took the osmotic pressure and its relation to the theory of diffusion and to the molecular theory of heat as the starting point of his analysis of Brownian motion. He writes in the paper that follows: "According to this theory, a dissolved molecule differs from a suspended body in size, *only* and it is difficult to see why suspended bodies should not produce the same osmotic pressure as an equal number of dissolved molecules" (see pp. 86–87).

On the other hand, Einstein pointed out, according to the "classical theory of thermodynamics," suspended particles— as macroscopic objects—should not exert an osmotic pressure on a semipermeable wall. Before Einstein, no one seems to have recognized that this contrast provides a touchstone for the kinetic theory. His choice of a suspension to study the relations between the thermodynamic and atomic theories of heat amounted to a radical reversal of perspective. Usually the legitimacy of microscopic explanations of thermodynamic results was at issue. In this case, however, the question centered on the applicability of a thermodynamic concept—osmotic pressure—to the suspended particles.

In the course of studying colloidal solutions, the commonly made distinction between suspensions and solutions in nineteenth-century chemistry had lost its absolute character. The absence of any fundamental difference between solutions and suspensions was made strikingly clear in 1902, when observations performed with the newly invented ultramicroscope made it possible to resolve many colloidal solutions into their constituents. The ultramicroscope not only demonstrated the physical reality of colloidal particles, but showed that irregular motion is one of their outstanding characteristics.

Although the ultramicroscope brought closer what Jean Perrin called the "distant reality" of molecules, one of their fundamental properties, their velocities, remained inaccessible to measurement. The inconsistencies that result from presumed velocity measurements, such as Exner's, had hinted at this problem; but it was explicitly discussed for the first time in the theoretical studies of Brownian motion that Einstein and Smoluchowski independently published between 1905 and 1907. Both introduced the mean-square displacement of the suspended particles as the primary observable quantity in Brownian motion. Einstein argued that dissipative forces change the direction and magnitude of the velocity of a suspended particle on such a short timescale that it cannot be measured. This argument demonstrates the fundamental role of dissipation in Einstein's analysis of Brownian motion.

In summary, the study of previous explanations of Brownian motion shows that three elements of Einstein's approach are characteristic of his decisive progress: (1) he based his analysis on the osmotic pressure rather than on the equipartition theorem; (2) he identified the mean-square displacements of suspended particles rather than their velocities as suitable observable quantities; and (3) he simultaneously applied the molecular theory of heat and the macroscopic theory of dissipation to the same phenomenon, rather than restricting each of these conceptual tools to a single scale, molecular or macroscopic.

IN THE PAPER on Brownian motion (paper 2), Einstein proved "that, on the assumption of the molecular theory of heat, bodies of the order of magnitude of 1/1000 mm suspended in liquids must already carry out an observable random movement, which is generated by thermal motion," as he wrote in a letter to Conrad Habicht in the late spring of 1905. Einstein wrote this paper "without knowing that observations concerning Brownian motion were already long familiar."[2] He did not mention Brownian motion in the title, although he conjectured that the motion he predicted might be identical to Brownian motion. Boltzmann's *Gastheorie*, which Einstein carefully studied during his student years, explicitly denies that the thermal motion of molecules in a gas leads to observable motions of suspended bodies. (This denial may be an instance of what Einstein referred to as Boltzmann's attaching too little importance to a comparison of theory with observation.) Sometime between 1902 and 1905, Einstein read Poincaré's *Science et hypothèse*, which contains a brief discussion of Gouy's work on Brownian motion, emphasizing Gouy's argument that Brownian motion violates the second law of thermodynamics. Einstein's second paper on Brownian motion,[3] written after Siedentopf drew his attention to Gouy's work, cites Gouy's observations as qualitative confirmation of his results.

Paper 2 opens with the derivation of an expression for the diffusion coefficient in terms of the radius of the suspended particles and the temperature and viscosity of the liquid, an expression already obtained in Einstein's doctoral dissertation. Unlike the previous derivation, however, the new one makes use of the methods of statistical physics that Einstein developed. The new approach is different in two respects:

1. In his dissertation, which deals with solutions rather than suspensions, Einstein simply assumed the validity of van't

Hoff's law for the osmotic pressure. He now gave a derivation of this law from an expression for the free energy of the suspension that follows from statistical mechanics.

2. Rather than simply considering the equilibrium of forces acting on a single molecule, Einstein derived the equilibrium between the osmotic pressure and a friction force obeying Stokes's law by a thermodynamical argument.

The ensuing derivation of the diffusion equation is based on the introduction of a probability distribution for displacements. The introduction of such a distribution is presumably related to Einstein's previous use of probability distributions. Einstein assumed the existence of a time interval, short with respect to the observation time, yet sufficiently long that the motions of a suspended particle in two successive time intervals can be treated as independent of each other. The displacement of the suspended particles can then be described by a probability distribution that determines the number of particles displaced by a certain distance in each time interval. Einstein derived the diffusion equation from an analysis of the time-dependence of the particle distribution, calculated from the probability distribution for displacements. This derivation is based on his crucial insight into the role of Brownian motion as the microscopic process responsible for diffusion on a macroscopic scale. Compared to such a derivation, one based on an analogy to the treatment of diffusion in the kinetic theory of gases may have appeared more problematic to Einstein because of the lack of a fully developed kinetic theory of liquids.

The solution of the resulting diffusion equation, combined with his expression for the diffusion coefficient, yields an expression for the mean-square displacement, λ_x, as a function of time, an expression that Einstein suggested could be used experimentally to determine Avogadro's number N:

$$\lambda_x = \sqrt{t}\sqrt{\frac{RT}{N}\frac{1}{3\pi kP}}, \tag{1}$$

where t is the time, R the gas constant, T the temperature, k the viscosity, and P the radius of the suspended particles.

Through his earlier work, Einstein was familiar with the theory of diffusion in both gases and liquids, as well as with other techniques needed for his analysis of Brownian motion. In 1902 he suggested the replacement of semipermeable walls in thermodynamic arguments by external conservative forces, a method he stated to be particularly useful for treating arbitrary mixtures. During 1903 Einstein discussed the notions of semipermeable membrane and osmotic pressure in his correspondence with Michele Besso, showing interest in Sutherland's hypothesis on the mechanism of semipermeable membranes. In his papers on statistical physics, Einstein generalized the idea of external conservative forces and noted the significant role of fluctuations in statistical physics. In 1904 he derived an expression for mean square deviations from the average value of the energy of a system.

SEVERAL newly perfected techniques for the experimental investigation of Brownian motion, notably the ultramicroscope and new methods for preparing colloidal solutions, were available by the time Einstein published his first articles on the subject. As one of the first applications of the ultramicroscope, Siedentopf and Zsigmondy observed Brownian motion in colloidal solutions but did not perform precise measurements. *Les ultramicroscopes et les objets ultramicroscopiques*, a book on the ultramicroscope and its applications by Aimé Cotton and Henri Mouton, published in 1906, helped stimulate interest in Brownian motion and brought Einstein's theory to the attention of researchers in the field. Using an ultramicroscope and a sophisticated observational technique, The Svedberg carried out careful measurements of Brownian motion with the aim of testing the interpretation of Brownian motion that maintains it is caused by the thermal motions of molecules. Svedberg reported on his attempts to test Einstein's theory in 1906 and corresponded with Einstein on the subject of Brownian motion, sending him one of his papers.

Svedberg followed Zsigmondy in assuming two types of motions for colloidal particles, a translational motion and a "proper [Brownian] motion." Svedberg restricted his attention to the latter and attempted to facilitate its measurement by superimposing a translational motion. He described the resulting trajectories as "sinusoid-like," but cautioned against concluding that the motion had an oscillatory character.[4] In the analysis of his results, however, Svedberg introduced a terminology that was adapted to the description of a simple oscillatory motion, relating the observed amplitudes to Einstein's root-mean-square displacement. Earlier, he had tried to estimate molecular velocities on the basis of the observed velocities of colloidal particles. In another article, written mainly to correct the basic misunderstandings in Svedberg's work, Einstein showed that the velocities of ultramicroscopic particles, as calculated from the equipartition theorem, cannot be directly observed.[5] On 11 November 1909, Einstein wrote to Perrin: "The errors in Svedberg's method of observation and also in his theoretical treatment became clear to me at once. I wrote a minor correction at the time, which only addressed the worst ones, as I couldn't bring myself to detract from Mr. S.'s great pleasure in his work."

In addition to the misconceptions underlying Svedberg's experimental work, his numerical results slightly disagreed with Einstein's predictions. Other early experimental work on Einstein's and Smoluchowski's theories, such as Felix Ehrenhaft's observations of displacements of aerosol particles, Victor Henri's cinematographical measurements of displacements of suspended particles, or Max Seddig's study of the temperature dependence of Brownian motion, provided qualitative confirmation of the theory; but the work of Henri and Seddig also failed to yield quantitative agreement. As a consequence, the kinetic interpretation of Brownian motion was not universally accepted in 1908 as the exclusive explanation of the phenomenon.

Einstein noted that the control of the temperature was the principal difficulty in obtaining satisfactory results from the photographic records of Seddig and Henri. Before Jean Perrin published his thorough experimental investigation of the phenomenon, Einstein was skeptical about the possibility of obtaining precise measurements of Brownian motion. On 30 July 1908, in a letter to Jakob Laub, Einstein commented rather enthusiastically on Seddig's work in spite of its shortcomings: "I have read Seddig's paper. He has done it very well. I cannot quite make head or tail of his descriptions of the results." Einstein wrote to Jean Perrin on 11 November the following year: "I would have considered it impossible to investigate Brownian motion so precisely; it is a stroke of luck for this subject that you have taken it up."

In a series of experiments, the first results of which were published in 1908, Perrin achieved an until then unmatched precision in the confirmation of almost all of Einstein's predictions. Like Einstein, Perrin recognized that the analogy established by van't Hoff between an ideal gas and a solution could be extended to colloidal solutions and suspensions, and that this analogy provides a unique means of obtaining evidence for atomism. In his first experiments on Brownian motion, Perrin tested a formula for the vertical distribution of suspended particles under the influence of gravitation. Although Perrin probably was aware of Einstein's theory through Langevin, he was apparently unaware that Einstein had derived a similar formula. Challenged by criticism, Perrin checked his assumption of the validity of Stokes's formula for the particles used in his experiments. In two further papers published in 1908, Perrin applied his methods to a determination of Avogadro's number.

In the same year, Perrin's doctoral student Chaudesaigues subjected Einstein's displacement formula to experimental tests. Contrary to Henri's results mentioned above, the results are in excellent agreement with theoretical predictions. Perrin continued these successful experiments with the help of other students; to Einstein's surprise, he was able to include rotational Brownian motion in his investigations. On 11 November 1909, Einstein wrote to Perrin: "I would not have considered a measurement of the rotations as feasible. In my eyes it was only a pretty trifle." Perrin's success was based on the ingenious combination of several experimental techniques for preparing emulsions with precisely controllable particle sizes, and for measuring the particles' number and displacements. He summarized his results in various review articles and books that significantly furthered the general acceptance of atomism.

Beginning in 1907, Einstein himself tried to contribute to the experimental study of fluctuation phenomena. His prediction of voltage fluctuations in condensers stimulated him to explore the possibility of measuring small quantities of electricity in order to provide experimental support for "a phenomenon in the field of electricity related to Brownian motion."[6] On 15 July 1907, he wrote to his friends Conrad and Paul Habicht about his discovery of a method for the measurement of small quantities of electrical energy. Soon afterward the Habichts tried to build the device proposed by Einstein. At the end of 1907, Einstein dropped his idea of obtaining a patent for the device, as he wrote on 24 December to Conrad Habicht, "primarily because of the lack of interest by manufacturers." Instead, he published a paper on the basic features of his method, which stimulated further work on the device.[7] While the use of the device for measuring fluctuation phenomena in conductors proved to be difficult, experimental work done by others soon provided evidence for the atomistic constitution of matter and electricity that exceeded Einstein's initial expectations.

EDITORIAL NOTES

[1] Einstein, *Autobiographical Notes*, Paul Arthur Schilpp, trans. and ed. (La Salle, Ill.: Open Court, 1979), pp. 44–45.

[2] Ibid.

[3] *Annalen der Physik* 19 (1906): 371–381, reprinted in *Collected Papers*, vol. 2, doc. 32, pp. 334–344.

[4] Svedberg, *Zeitschrift für Elektrochemie und angewandte physikalische Chemie* 12 (1906): 853–854.

[5] Einstein, in ibid., 13 (1907): 41–42; reprinted in *Collected Papers*, vol. 2, doc. 40, pp. 399–400.

[6] *Annalen der Physik* 22 (1907): 569–572, reprinted in *Collected Papers*, vol. 2, doc. 39, pp. 393–396.

[7] *Physikalische Zeitschrift* 9 (1908): 216–217, reprinted in *Collected Papers*, vol. 2, doc. 48, pp. 490–491.

PAPER 2

✦

On the Motion of Small Particles Suspended in Liquids at Rest Required by the Molecular-Kinetic Theory of Heat

IN THIS PAPER it will be shown that, according to the molecular-kinetic theory of heat, bodies of a microscopically visible size suspended in liquids must, as a result of thermal molecular motions, perform motions of such magnitude that they can be easily observed with a microscope. It is possible that the motions to be discussed here are identical with so-called Brownian molecular motion; however, the data available to me on the latter are so imprecise that I could not form a judgment on the question.

If the motion to be discussed here can actually be observed, together with the laws it is expected to obey, then classical thermodynamics can no longer be viewed as applying to regions that can be distinguished even with a microscope, and an exact determination of actual atomic sizes becomes possible. On the other hand, if the prediction of this motion were to be proved wrong, this fact would provide a far-reaching argument against the molecular-kinetic conception of heat.

1. On the Osmotic Pressure to Be Ascribed to Suspended Particles

Let z gram-molecules of a non-electrolyte be dissolved in a part V^* of the total volume V of a liquid. If the volume V^* is separated from the pure solvent by a wall that is permeable to the solvent but not to the solute, then this wall is subjected to a so-called osmotic pressure, which for sufficiently large values of V^*/z satisfies the equation

$$pV^* = RTz.$$

But if instead of the solute, the partial volume V^* of the liquid contains small suspended bodies that also cannot pass through the solvent-permeable wall, then, according to the classical theory of thermodynamics, we should not expect— at least if we neglect the force of gravity, which does not interest us here—any pressure to be exerted on the wall; for according to the usual interpretation, the "free energy" of the system does not seem to depend on the position of the wall and of the suspended bodies, but only on the total mass and properties of the suspended substance, the liquid, and the wall, as well as on the pressure and temperature. To be sure, the energy and entropy of the interfaces (capillary forces) should also be considered when calculating the free energy; but we can disregard them here because changes in the position of the wall and suspended bodies will not cause changes in the size and state of the contact surfaces.

But a different interpretation arises from the standpoint of the molecular-kinetic theory of heat. According to this

theory, a dissolved molecule differs from a suspended body *only* in size, and it is difficult to see why suspended bodies should not produce the same osmotic pressure as an equal number of dissolved molecules. We have to assume that the suspended bodies perform an irregular, albeit very slow, motion in the liquid due to the liquid's molecular motion; if prevented by the wall from leaving the volume V^*, they will exert pressure upon the wall just like molecules in solution. Thus, if n suspended bodies are present in the volume V^*, i.e., $n/V^* = \nu$ in a unit volume, and if neighboring bodies are sufficiently far separated from each other, there will be a corresponding osmotic pressure p of magnitude

$$p = \frac{RT}{V^*}\frac{n}{N} = \frac{RT}{N} \cdot \nu,$$

where N denotes the number of actual molecules per gram-molecule. It will be shown in the next section that the molecular-kinetic theory of heat does indeed lead to this broader interpretation of osmotic pressure.

2. Osmotic Pressure from the Standpoint of the Molecular-Kinetic Theory of Heat[1]

If $p_1 p_2 \ldots p_l$ are state variables of a physical system that completely determine the system's instantaneous state (e.g., the coordinates and velocity components of all atoms of the

[1] In this section it is assumed that the reader is familiar with the author's papers on the foundations of thermodynamics (cf. *Ann. d. Phys.* 9 [1902]: 417 and 11 [1903]: 170). Knowledge of these papers and of this section of the present paper is not essential for an understanding of the results in the present paper.

system), and if the complete system of equations for changes of these variables is given in the form

$$\frac{\partial p_\nu}{\partial t} = \varphi_\nu(p_1 \dots p_l) \quad (\nu = 1, 2, \dots l),$$

with $\sum \frac{\partial \varphi_\nu}{\partial p_\nu} = 0$, then the entropy of the system is given by the expression

$$S = \frac{\bar{E}}{T} + 2\kappa \ln \int e^{-\frac{E}{2\kappa T}} dp_1 \dots dp_l.$$

Here T denotes the absolute temperature, $\bar{E}$ the energy of the system, and E the energy as a function of the p_ν. The integral extends over all possible values of p_ν consistent with the conditions of the problem. κ is connected with the constant N mentioned above by the relation $2\kappa N = R$. Hence we get for the free energy F

$$F = -\frac{R}{N} T \ln \int e^{-\frac{EN}{RT}} dp_1 \dots dp_l = -\frac{RT}{N} \ln B.$$

Let us now imagine a liquid enclosed in volume V; let a part V^* of the volume V contain n solute molecules or suspended bodies, which are retained in the volume V^* by a semipermeable wall; the integration limits of the integral B occurring in the expressions for S and F will be affected accordingly. Let the total volume of the solute molecules or suspended bodies be small compared with V^*. In accordance with the theory mentioned, let this system be completely described by the variables $p_1 \dots p_l$.

Even if the molecular picture were extended to include *all* details, the calculation of the integral B would be so difficult that an exact calculation of F is hardly conceivable. However, here we only need to know how F depends on the size of the volume V^* in which all the solute molecules

or suspended bodies (hereafter called "particles" for brevity) are contained.

Let us call the rectangular coordinates of the center of gravity of the first particle x_1, y_1, z_1, those of the second x_2, y_2, z_2, etc., and those of the last particle x_n, y_n, z_n, and assign to the centers of gravity of the particles the infinitesimally small parallelepiped regions $dx_1 dy_1 dz_1$, $dx_2 dy_2 dz_2 \ldots dx_n dy_n dz_n$, all of which lie in V^*. We want to evaluate the integral occurring in the expression for F, with the restriction that the centers of gravity of the particles shall lie in the regions just assigned to them. In any case, this integral can be put into the form

$$dB = dx_1 dy_1 \ldots dz_n \cdot J,$$

where J is independent of $dx_1 dy_1$, etc., as well as of V^*, i.e., of the position of the semipermeable wall. But J is also independent of the particular choice of the *positions* of the center of gravity regions and of the value of V^*, as will be shown immediately. For if a second system of infinitesimally small regions were assigned to the centers of gravity of the particles and denoted by $dx_1' dy_1' dz_1'$, $dx_2' dy_2' dz_2' \ldots dx_n' dy_n' dz_n'$, and if these regions differed from the originally assigned ones by their position alone, but not by their size, and if, likewise, all of them were contained in V^*, we would similarly have

$$dB' = dx_1' dy_1' \ldots dz_n' \cdot J',$$

where

$$dx_1 dy_1 \ldots dz_n = dx_1' dy_1' \ldots dz_n'.$$

Hence,

$$\frac{dB}{dB'} = \frac{J}{J'}.$$

But from the molecular theory of heat, as presented in the papers cited,[2] it is easily deduced that dB/B and dB'/B are respectively equal to the probabilities that at an arbitrarily chosen moment the centers of gravity of the particles will be found in the regions $(dx_1 \ldots dz_n)$ and $(dx_1' \ldots dz_n')$ respectively. If the motions of the individual particles are independent of one another (to a sufficient approximation) and if the liquid is homogeneous and no forces act on the particles, then for regions of the same size the probabilities of the two systems of regions will be equal, so that

$$\frac{dB}{B} = \frac{dB'}{B}.$$

But from this equation and the previous one it follows that

$$J = J'.$$

This proves that J does not depend on either V^* or $x_1, y_1 \ldots z_n$. By integrating, we get

$$B = \int J \, dx_1 \ldots dz_n = JV^{*n},$$

and hence

$$F = -\frac{RT}{N}\{\ln J + n \, \ln V^*\}$$

and

$$p = -\frac{\partial F}{\partial V^*} = \frac{RT}{V^*}\frac{n}{N} = \frac{RT}{N}\nu.$$

This analysis shows that the existence of osmotic pressure can be deduced from the molecular-kinetic theory of heat, and that, at high dilution, according to this theory, equal

[2] A. Einstein, *Ann. d. Phys.* 11 (1903): 170.

numbers of solute molecules and suspended particles behave identically as regards osmotic pressure.

3. THEORY OF DIFFUSION OF SMALL SUSPENDED SPHERES

Suppose suspended particles are randomly distributed in a liquid. We will investigate their state of dynamic equilibrium under the assumption that a force K, which depends on the position but not on the time, acts on the individual particles. For the sake of simplicity, we shall assume that the force acts everywhere in the direction of the X-axis.

If the number of suspended particles per unit volume is ν, then in the case of thermodynamic equilibrium ν is a function of x such that the variation of the free energy vanishes for an arbitrary virtual displacement δx of the suspended substance. Thus

$$\delta F = \delta E - T\, \delta S = 0.$$

Let us assume that the liquid has a unit cross section perpendicular to the X-axis, and that it is bounded by the planes $x = 0$ and $x = l$. We then have

$$\delta E = -\int_0^l K\nu\, \delta x\, dx$$

and

$$\delta S = \int_0^l R\frac{\nu}{N}\frac{\partial \delta x}{\partial x}\, dx = -\frac{R}{N}\int_0^l \frac{\partial \nu}{\partial x}\, \delta x\, dx.$$

The required equilibrium condition is therefore

$$-K\nu + \frac{RT}{N}\frac{\partial \nu}{\partial x} = 0 \tag{1}$$

91

or

$$Kv - \frac{\partial p}{\partial x} = 0.$$

The last equation asserts that the force K is equilibrated by the force of osmotic pressure.

We can use equation (1) to determine the diffusion coefficient of the suspended substance. We can look upon the dynamic equilibrium state considered here as a superposition of two processes proceeding in opposite directions, namely:

1. A motion of the suspended substance under the influence of the force K that acts on each suspended particle.
2. A process of diffusion, which is to be regarded as the result of the disordered motions of the particles produced by thermal molecular motion.

If the suspended particles have spherical form (where P is the radius of the sphere) and the coefficient of viscosity of the liquid is k, then the force K imparts to an individual particle the velocity[3]

$$\frac{K}{6\pi kP},$$

and

$$\frac{vK}{6\pi kP}$$

particles will pass through a unit area per unit time.

Further, if D denotes the diffusion coefficient of the suspended substance and μ the mass of a particle, then

$$-D\frac{\partial(\mu v)}{\partial x} \text{ grams}$$

[3] Cf., e.g., G. Kirchhoff, *Vorlesungen über Mechanik*, 26. Vorl., §4 (*Lectures on Mechanics*, Lecture 26, sec. 4).

or

$$-D\frac{\partial \nu}{\partial x}$$

particles will pass across a unit area per unit time as the result of diffusion. Since dynamic equilibrium prevails, we must have

$$\frac{\nu K}{6\pi kP} - D\frac{\partial \nu}{\partial x} = 0. \tag{2}$$

From the two conditions (1) and (2) found for dynamic equilibrium, we can calculate the diffusion coefficient. We get

$$D = \frac{RT}{N} \cdot \frac{1}{6\pi kP}.$$

Thus, except for universal constants and the absolute temperature, the diffusion coefficient of the suspended substance depends only on the viscosity of the liquid and on the size of the suspended particles.

4. On the Disordered Motion of Particles Suspended in a Liquid and Its Relation to Diffusion

We shall now turn to a closer examination of the disordered motions that arise from thermal molecular motion and give rise to the diffusion investigated in the last section.

Obviously, we must assume that each individual particle executes a motion that is independent of the motions of all the other particles; the motions of the same particle in different time intervals must also be considered as mutually independent processes, so long as we think of these time intervals as chosen not to be too small.

We now introduce a time interval τ, which is very small compared with observable time intervals but still large enough that the motions performed by a particle during two consecutive time intervals τ can be considered as mutually independent events.

Suppose, now, that a total of n suspended particles is present in a liquid. In a time interval τ, the X-coordinates of the individual particles will increase by Δ, where Δ has a different (positive or negative) value for each particle. A certain probability distribution law will hold for Δ: the number dn of particles experiencing a displacement that lies between Δ and $\Delta + d\Delta$ in the time interval τ will be expressed by an equation of the form

$$dn = n\varphi(\Delta)\,d\Delta,$$

where

$$\int_{-\infty}^{+\infty} \varphi(\Delta)\,d\Delta = 1,$$

and φ differs from zero only for very small values of Δ and satisfies the condition

$$\varphi(\Delta) = \varphi(-\Delta).$$

We will now investigate how the diffusion coefficient depends on φ, restricting ourselves again to the case where the number ν of particles per unit volume only depends on x and t.

Let $\nu = f(x, t)$ be the number of particles per unit volume; we calculate the distribution of particles at time $t + \tau$ from their distribution at time t. From the definition of the function $\varphi(\Delta)$ we can easily obtain the number of particles found at time $t + \tau$ between two planes perpendicular to the X-axis with abscissas x and $x + dx$. We get

$$f(x, t + \tau)\, dx = dx \cdot \int_{\Delta = -\infty}^{\Delta = +\infty} f(x + \Delta)\varphi(\Delta)\, d\Delta. \text{[1]}$$

But since τ is very small, we can put

$$f(x, t + \tau) = f(x, t) + \tau \frac{\partial f}{\partial t}.$$

Further, let us expand $f(x + \Delta, t)$ in powers of Δ :

$$f(x + \Delta, t) = f(x, t) + \Delta \, \frac{\partial f(x, t)}{\partial x} + \frac{\Delta^2}{2!} \, \frac{\partial^2 f(x, t)}{\partial x^2} \, \ldots \text{ ad inf.}$$

We can bring this expansion under the integral sign since only very small values of Δ contribute anything to the latter. We obtain

$$f + \frac{\partial f}{\partial t} \cdot \tau = f \cdot \int_{-\infty}^{+\infty} \varphi(\Delta)\, d\Delta + \frac{\partial f}{\partial x} \int_{-\infty}^{+\infty} \Delta \varphi(\Delta)\, d\Delta$$
$$+ \frac{\partial^2 f}{\partial x^2} \int_{-\infty}^{+\infty} \frac{\Delta^2}{2} \varphi(\Delta)\, d\Delta \ldots$$

On the right-hand side, the second, fourth, etc., terms vanish since $\varphi(x) = \varphi(-x)$, while for the first, third, fifth, etc., terms, each successive term is very small compared with the one preceding it. From this equation, by taking into account that

$$\int_{-\infty}^{+\infty} \varphi(\Delta)\, d\Delta = 1,$$

and putting

$$\frac{1}{\tau} \int_{-\infty}^{+\infty} \frac{\Delta^2}{2} \varphi(\Delta)\, d\Delta = D,$$

and taking into account only the first and third terms of the right-hand side, we get

$$\frac{\partial f}{\partial t} = D \frac{\partial^2 f}{\partial x^2}. \tag{3}$$

This is the well-known differential equation for diffusion, and we recognize that D is the diffusion coefficient.

Another important point can be linked to this argument. We have assumed that the individual particles are all referred to the same coordinate system. However, this is not necessary since the motions of the individual particles are mutually independent. We will now refer the motion of each particle to a coordinate system whose origin coincides with the position of the center of gravity of the particle in question at time $t = 0$, with the difference that $f(x, t)dx$ now denotes the number of particles whose X-coordinate has *increased* between the times $t = 0$ and $t = t$ by a quantity that lies somewhere between x and $x + dx$. Thus, the function f varies according to equation (1) in this case as well.[2] Further, it is obvious that for $x \gtrless 0$ and $t = 0$ we must have

$$f(x, t) = 0 \text{ and } \int_{-\infty}^{+\infty} f(x, t)\, dx = n.$$

The problem, which coincides with the problem of diffusion outwards from a point (neglecting the interaction between the diffusing particles), is now completely determined mathematically; its solution is

$$f(x, t) = \frac{n}{\sqrt{4\pi D}} \frac{e^{-\frac{x^2}{4Dt}}}{\sqrt{t}}.$$

The probability distribution of the resulting displacements during an arbitrary time t is thus the same as the distribution of random errors, which was to be expected. What is important, however, is how the constant in the exponent is related to the diffusion coefficient. With the help of this equation we can now calculate the displacement λ_x in the direction of the X-axis that a particle experiences on the average, or, to be more precise, the root-mean-square displacement in the X-direction; it is

$$\lambda_x = \sqrt{\overline{x^2}} = \sqrt{2Dt}.$$

The mean displacement is thus proportional to the square root of the time. It can easily be shown that the root mean square of the *total displacements* of the particles has the value $\lambda_x \sqrt{3}$.

5. Formula for the Mean Displacement of Suspended Particles. A New Method of Determining the Actual Size of Atoms

In section 3 we found the following value for the diffusion coefficient D of a substance suspended in a liquid in the form of small spheres of radius P :

$$D = \frac{RT}{N} \frac{1}{6\pi kP}.$$

Further, we found in section 4 that the mean value of the displacements of the particles in the X-direction at time t equals

$$\lambda_x = \sqrt{2Dt}.$$

By eliminating D, we get:

$$\lambda_x = \sqrt{t} \cdot \sqrt{\frac{RT}{N} \frac{1}{3\pi kP}}.$$

This equation shows how λ_x depends on T, k, and P.

We will now calculate how large λ_x is for one second if N is taken to be $6 \cdot 10^{23}$ in accordance with the results of the kinetic theory of gases; water at 17°C ($k = 1.35 \cdot 10^{-2}$) is chosen as the liquid,[3] and the diameter of the particles is 0.001 mm. We get

$$\lambda_x = 8 \cdot 10^{-5} \text{ cm} = 0.8 \text{ micron.}$$

Therefore, the mean displacement in one minute would be about 6 microns.

Conversely, the relation can be used to determine N. We obtain

$$N = \frac{t}{\lambda_x^2} \cdot \frac{RT}{3\pi kP}.$$

Let us hope that a researcher will soon succeed in solving the problem presented here, which is so important for the theory of heat.

(Annalen der Physik 17 [1905]: 549–560)

EDITORIAL NOTES

[1] f on the right-hand side is to be taken at the time t.
[2] Here eq. (1) should be eq. (3).
[3] The value of the viscosity of water is taken from paper 1, p. 65, and actually refers to water at temperature 9.5°C.

Part Three

Einstein on the Theory of Relativity

oder

$$k \, y = \frac{d}{dt} \left\{ \frac{mc^2}{\sqrt{1-\frac{q^2}{c^2}}} \right\} . \quad \ldots \quad (27a)$$

Der Ausdruck unter der Klammer rechts spi~~
~~Energie~~ *Edes bewegten Massenpunktes)* wobei allerdings ~~war eine Integration~~ nächst für

$$\mathcal{E} = \frac{mc^2}{\sqrt{1-\frac{q^2}{c^2}}} \quad \ldots \quad \text{~~}(28)$$

wächst ins Unendliche, wenn sich q dem Wer~
also einer unendlichen Energie-Aufwendes, ~
Geschwindigkeit c zu erteilen. Um zu sehen, dass ~
kleine Geschwindigkeit in den von Newtons Me~
entwickeln wir den Nenner *nach Potenzen von $\frac{q^2}{c^2}$* und erhalten

$$\mathcal{E} = mc^2 + \frac{m}{2}q^2 + \cdots \quad (28')$$

Das zweite Glied ~~~~ der rechten Seite ist der gel~

Part of Einstein's 1912 manuscript on the special theory of relativity, showing the equation $E = mc^2$. (Courtesy of Hebrew University of Jerusalem)

Einstein was the first physicist to formulate clearly the new kinematical foundation for all of physics inherent in Lorentz's electron theory. This kinematics emerged in 1905 from his critical examination of the physical significance of the concepts of spatial and temporal intervals. The examination, based on a careful definition of the simultaneity of distant events, showed that the concept of a universal or absolute time, on which Newtonian kinematics is based, has to be abandoned; and that the Galilean transformations between the coordinates of two inertial frames of reference has to be replaced by a set of spatial and temporal transformations that agree formally with a set that Lorentz had introduced earlier with a quite different interpretation. Through the interpretation of these transformations as elements of a space-time symmetry group corresponding to the new kinematics, the special theory of relativity (as it later came to be called) provided physicists with a powerful guide in the search for new dynamical theories of fields and particles and gradually led to a deeper appreciation of the role of symmetry criteria in physics. The special theory of relativity also provided philosophers with abundant material for reflection on the new views of space and time. The special theory, like Newtonian mechanics, still assigns a privileged status to the class of inertial frames of reference. The attempt to generalize the theory to include gravitation led Einstein to formulate the equivalence principle in 1907. This was the first step in his search for a new theory of gravitation denying a privileged role to inertial frames, a theory that is now known as the general theory of relativity.

Einstein presented the special theory in paper 3, which is a landmark in the development of modern physics. In the first part of this paper Einstein presented the new kinematics, basing it on two postulates, the relativity principle and the principle of the constancy of the velocity of light. In the second part, he applied his kinematical results to the solution of a number of problems in the optics and electrodynamics of moving bodies. In paper 4, Einstein presents arguments for one of the most important consequences of the theory, the equivalence of mass and energy.

Strictly speaking, it is anachronistic to use the term "theory of relativity" in discussing Einstein's first papers on the subject. In them he referred to the "principle of relativity." Max Planck used the term "Relativtheorie" in 1906 to describe the Lorentz-Einstein equations of motion for the electron, and this expression continued to be used from time to time for several years. A. H. Bucherer seems to have been the first person to use the term "Relativitätstheorie" in the discussion following Planck's lecture. The term was used in an article by Paul Ehrenfest and adopted by Einstein in 1907 in his reply to this article. Although Einstein used the term from time to time thereafter, for several years he continued to employ "Relativitätsprinzip" in the titles of his articles. In 1910 the mathematician Felix Klein suggested the name "Invariantentheorie" (theory of invariants), but this suggestion does not seem to have been adopted by any physicist. In 1915 Einstein started to refer to his earlier work as "the special theory of relativity" to contrast it with his later "general theory."

IN HIS 1905 paper, as well as in his 1907 and 1909 reviews of the theory, Einstein described the theory of relativity as arising from a specific problem: the apparent conflict between the principle of relativity and the Maxwell-Lorentz theory of electrodynamics. While the relativity principle asserts the physical equivalence of all inertial frames of reference, the Maxwell-Lorentz theory implies the existence of a privileged inertial frame.

The principle of relativity originated in classical mechanics. Assuming Newton's laws of motion and central force interactions, it can be demonstrated that it is impossible to determine the state of motion of an inertial frame by means of mechanical experiments carried out within a closed system with center of mass at rest in this frame. This conclusion, well known and empirically well confirmed by the end of the nineteenth century, was sometimes called the principle of relative motion, or principle of relativity.

The introduction of velocity-dependent forces between charged particles led to doubts about the validity of the relativity principle for magnetic interactions. The wave theory of light appeared to invalidate the principle for optical phenomena. This theory seemed to require an all-pervading medium, the so-called luminiferous ether, to explain the propagation of light in the absence of ordinary matter. The assumption that the ether moves together with matter seems to be excluded by the phenomenon of aberration and by Fizeau's results on the velocity of light in moving media. If the ether is not dragged with matter, it should be possible to detect motion relative to a reference frame fixed in the ether by means of optical experiments. However, all attempts to detect the motion of the earth through the ether by optical experiments failed.

Maxwell's electromagnetic theory was intended to provide a unified explanation of electric, magnetic, and optical phenomena. With its advent, the question arose of the status of the principle of relativity for such phenomena. Does the principle follow from the fundamental equations of electrodynamics? The answer to this question depends on the form of Maxwell's equations postulated for bodies in motion. Hertz developed an electrodynamics of moving bodies, based on the assumption that the ether moves with matter, in which the relativity principle holds. In addition to its inability to account for the optical phenomena mentioned above, Hertz's theory was unable to explain several new electromagnetic phenomena, and it soon fell out of favor.

By the turn of this century, when Einstein started working on the electrodynamics of moving bodies, Lorentz's very successful version of Maxwell's theory had gained wide acceptance. Lorentz's electrodynamics is based on a microscopic theory that came to be known as the electron theory. The theory makes a sharp distinction between ordinary, ponderable matter and the ether. Ordinary matter is composed of finite-sized material particles, at least some of which are electrically charged. All of space, even those regions occupied by material particles, is pervaded by the ether, a medium with no mechanical properties, such as mass. The ether is the seat of all electric and magnetic fields. Matter only influences the ether through charged particles, which create these fields. The ether only acts on matter through the electric and magnetic forces that the fields exert on charged particles. By assuming such atoms of electricity ("electrons"), Lorentz's theory incorporates an important element of the pre-Maxwellian continental tradition into Maxwell's theory, from which it took the field equations.

The parts of the ether are assumed to be immobile relative to each other. Hence, Lorentz's ether defines a rigid reference frame, which is assumed to be inertial. It is in this frame that Maxwell's equations are valid; in other frames, the Galilei-transformed form of these equations hold. Hence it should be possible to detect the motion of the earth through the ether by suitably designed terrestrial electromagnetic or optical experiments. Lorentz was well aware of the failure of all attempts to detect the motion of the earth through the ether, in particular such sensitive optical attempts as the Michelson-Morley experiment, and attempted to explain this failure on the basis of his theory.

His basic approach to this problem in 1895 was to use the theorem of "corresponding states" in combination with the well-known contraction hypothesis. The theorem is essentially a calculational tool that sets up a correspondence between phenomena in moving systems and those in stationary systems by introducing transformed coordinates and fields. On this basis, Lorentz was able to account for the failure of most electromagnetic experiments to detect the motion of the earth through the ether. In 1904 he showed how to explain the failure of all such experiments by a generalization of his theorem. He introduced a set of transformations for the spatial and temporal coordinates (soon named the "Lorentz transformations" by Poincaré) and for the electric and magnetic field components, such that by using these transformations, Maxwell's equations, in the absence of charges, take the same form in all inertial frames. Lorentz's approach to the explanation of the failure of attempts to detect motion through the ether, thus, was to show that the basic equations of the electron theory, in spite of the fact that they single out the ether rest frame, can still explain this failure of all optical and electromagnetic attempts to detect the earth's motion through the ether.

Einstein's work was based on a new outlook on the problem. Instead of regarding the failure of electromagnetic and optical experiments to detect the earth's motion through the ether as something to be deduced from the electrodynamical equations, he took this failure as empirical evidence for the validity of the principle of relativity in electrodynamics and optics. Indeed, he asserted the universal validity of the principle, making it a criterion for the acceptability of any physical law. In this respect he gave the principle of relativity a role similar to that of the principles of thermodynamics, an example that he later stated helped to guide him. Rather than being deductions from other theories, such principles are taken as postulates for deductive chains of reasoning, resulting in the formulation of general criteria that all physical theories must satisfy.

Einstein now confronted the problem of making Maxwell-Lorentz electrodynamics compatible with the principle of relativity. He did so by means of a principle drawn from this very electrodynamics, the principle of the constancy of the velocity of light. That the velocity of light is independent of that of its source, and has a constant value in the ether rest frame, can be deduced from the Maxwell-Lorentz theory. Einstein dropped the ether from that theory and took the constancy of the velocity of light as a second postulate, supported by all the empirical evidence in favor of the Maxwell-Lorentz theory. When combined with the relativity principle, this leads to an apparently paradoxical conclusion: the velocity of light must be the same in all inertial frames. This result conflicts with the Newtonian law of addition of velocities, forcing a revision of the kinematical foundations underlying all of physics. Einstein showed that the simultaneity of distant events is only defined physically relative to a particular inertial frame, leading to kinematical transformations between the spatial and temporal coordinates of two inertial frames that agree formally with the transformations that Lorentz had introduced in 1904.

Einstein next considered the implications of the new kinematics for electrodynamics and mechanics. By eliminating the concept of the ether, he in effect asserted that electromagnetic fields do not require an underlying substratum. He showed that the Maxwell-Lorentz equations for empty space remain invariant in form under the new kinematical transformations when the transformation laws for the electric and magnetic fields are appropriately defined. He deduced appropriate transformation laws for charge densities and velocities from the requirement that Maxwell's equations remain invariant when convection currents are added. Finally, by assuming that Newton's equations hold for a charged particle at rest, he was able to use a kinematical transformation to deduce the equations of motion of a charged particle ("electron") with arbitrary velocity.

The problems connected with the formulation of an electrodynamics of moving bodies consistent with all experimental evidence were discussed frequently during the years Einstein was working on his theory. Statements similar to many of the individual points made in paper 3 occur in the contemporary literature, and Einstein may well have been familiar with some of the books and articles in which they are found. But his approach to the problem, leading to the peculiar combination of these ideas in his paper, is unique— particularly the recognition that a new kinematics of universal applicability is needed as the basis for a consistent approach to the electrodynamics of moving bodies.

EINSTEIN'S work on relativity grew out of his long-standing interest in the electrodynamics and optics of moving bodies His first scientific essay, written in 1895, discussed the propagation of light through the ether. The next year, as he later recalled, the following problem started to puzzle him: "If one were to pursue a light wave with the velocity of light, one would be confronted with a time-independent wave field. Such a thing doesn't seem to exist, however! This was the first childlike thought-experiment concerned with the special theory of relativity."[1]

By this time Einstein presumably was familiar with the principle of relativity in classical mechanics. While preparing for the ETH entrance examination in 1895, he had studied the German edition of Violle's textbook. Violle actually based his treatment of dynamics on the "principle of relative motions" together with the principle of inertia.

About 1898, Einstein started to study Maxwell's electromagnetic theory, apparently with the help of Drude's textbook. By 1899, after studying Hertz's papers on the subject, he was at work on the electrodynamics of moving bodies. He discussed this topic a number of times in letters to Mileva Marić between 1899 and 1901; once, on 27 March 1901, he referred to "our work on relative motion." In December 1901, Einstein also explained his ideas on the subject to Professor Alfred Kleiner of the University of Zurich, who encouraged him to publish them; but there is no evidence that Kleiner played a further role in the development of these ideas.

Einstein's comments show that in 1899 his viewpoint on electrodynamics was similar to that of Lorentz; but, aside from this similarity, there is no evidence that Einstein had yet read anything by Lorentz. Shortly afterward, Einstein designed an experiment to test the effect of the motion of bodies relative to the ether on the propagation of light; in 1901 he designed a second such experiment, but was unable to carry out either one. On 17 December 1901, he reported to Marić that he was at work on "a capital paper" on the electrodynamics of moving bodies, asserting his renewed conviction of the correctness of his "ideas on relative motion." His words may indicate that he already doubted whether motion with respect to the ether is experimentally detectable. Soon afterward he wrote that he intended to study Lorentz's theory in earnest.

There is direct or strong indirect contemporary evidence that, by 1902, Einstein had read or was reading works on electrodynamics and optics by Drude, Helmholtz, Hertz, Lorentz, Voigt, Wien, and Föppl. Comments in his letters on articles published in the *Annalen der Physik* between 1898 and 1901 indicate that during those years he looked at that journal regularly as well, and studied a number of articles in it. It is reasonable to suppose that he continued to do so between 1902 and 1905. During these years a number of significant articles on the electrodynamics and optics of moving bodies appeared in the *Annalen*. He cited several works published before 1905 in his later articles on relativity, and it is possible that he read one or more of these before 1905. Einstein also read extensively on the foundations of science. He later attributed great significance for his development of the theory of relativity to his reading of Hume, Mach, and Poincaré.

Belief in the reality of the ether was widespread at the turn of the century. However, Einstein was familiar with several works that questioned the certainty of its existence. Mill, in the course of a discussion of "the Hypothetical Method" in his *Logic*, gives a number of reasons for skepticism concerning "the prevailing hypothesis of a luminiferous ether."[2] Poincaré, in *La science et l'hypothèse*, raised the question of the existence of the ether, even if he offered no clear answer. Ostwald, in his *Lehrbuch der allgemeinen Chemie*, suggested that the ether hypothesis could be replaced by a purely energetic treatment of radiation.

Few contemporary documents throw any light on Einstein's work on electrodynamics between 1902 and 1905. On 22 January 1903, he wrote Michele Besso: "In the near future I want to deal with molecular forces in gases, and then make a comprehensive study of electron theory." On 5 December 1903, Einstein gave a talk to the Naturforschende Gesellschaft Bern on "The Theory of Electromagnetic Waves." By the time he wrote his friend Conrad Habicht in May or June 1905, the theory was practically complete: "The . . . paper exists only as a sketch and is an electrodynamics of moving bodies that utilizes a modification of the theory of space and time."

Later reminiscences by Einstein suggest several important elements in the development of his ideas on relativity before "On the Electrodynamics of Moving Bodies" was written that are not recorded in any known contemporary documents. In a letter to Erika Oppenheimer on 13 September 1932, he gave a general characterization of "the situation that led to setting up the theory of special relativity": "Mechanically all inertial systems are equivalent. In accordance with experience, this equivalence also extends to optics and electrodynamics. However, it did not appear that this equivalence could be attained in the theory of the latter. I soon reached the conviction that this had its basis in a deep incompleteness of the theoretical system. The desire to discover and overcome this generated a state of psychic tension in me that, after seven years of vain searching, was resolved by relativizing the concepts of time and length."

In 1952 he wrote: "My direct path to the special theory of relativity was mainly determined by the conviction that the electromotive force induced in a conductor moving in a magnetic field is nothing other than an electric field. But the result of Fizeau's experiment and the phenomenon of aberration also guided me."[3]

Beyond their well-known role as evidence against the assumption that the ether is completely carried along by moving matter, it is not clear what role the result of Fizeau's experiment and the phenomenon of aberration played in Einstein's thinking. Possibly its role depended on the fact that, in both cases, the observed effect only depends on the motion of matter (water in the first case, a star in the second) relative to the earth, and not on the presumed motion of the earth with respect to the ether.

In the case of electromagnetic induction, Einstein gave a more detailed account of its role. In 1920, he wrote: "In setting up the special theory of relativity, the following . . . idea about Faraday's electromagnetic induction played a guiding role. According to Faraday, relative motion of a magnet and a closed electric circuit induces an electric current in the latter. Whether the magnet is moved or the conductor doesn't matter; only the relative motion is significant. . . . The phenomena of electromagnetic induction . . . compelled me to postulate the principle of (special) relativity." In a footnote he added: "The difficulty to be overcome then lay in the constancy of the velocity of light in vacuum, which I first thought would have to be abandoned. Only after groping for years did I realize that the difficulty lay in the arbitrariness of the fundamental concepts of kinematics."[4]

His strong belief in the relativity principle and abandonment of "the constancy of the velocity of light in vacuum" led Einstein to explore the possibility of an emission theory of light. In such a theory, the velocity of light is only constant relative to that of its source, so it is clearly consistent with the relativity principle. Newton's corpuscular theory of light is an emission theory, and Einstein's search for such a theory may have been connected with his light-quantum hypothesis (see paper 5). On 25 April 1912, in a letter to Paul Ehrenfest commenting on Ritz's emission theory, Einstein referred to "Ritz's conception, which before the theory of relativity was also mine." He expanded on this remark on 20 June: "I knew that the principle of the constancy of the velocity of light was something quite independent of the relativity postulate, and I weighed which was more probable, the principle of the constancy of c [the speed of light], as required by Maxwell's equations, or the constancy of c exclusively for an observer located at the light source. I decided in favor of the former."

In 1924, Einstein described the sudden resolution of his dilemma: "After seven years of reflection in vain (1898–1905), the solution came to me suddenly with the thought that our concepts and laws of space and time can only claim validity insofar as they stand in a clear relation to our experiences; and that experience could very well lead to the alteration of these concepts and laws. By a revision of the concept of simultaneity into a more malleable form, I thus arrived at the special theory of relativity."[5]

In a talk at Kyoto University in 1922, Einstein is reported to have said that, after a year of struggle with the problem of how to reconcile Lorentz's theory with his ideas on relativity, he visited a friend one day to discuss the problem in detail with him. The next day Einstein said to his friend: "Thanks to you, I have completely solved my problem."[6] The friend was presumably Michele Besso, then his colleague at the Swiss Patent Office and the only person whose help is acknowledged in paper 3.

Work on this paper was apparently completed very rapidly after this. In March 1952 Einstein wrote to Carl Seelig that "between the conception of the idea for the special theory of relativity and the completion of the relevant publication, five or six weeks elapsed."

Einstein's comments suggest the following stages in his work on the theory of relativity:

1. He became convinced that, as is the case for mechanical phenomena, only the relative motions of ponderable bodies are significant in determining electromagnetic and optical phenomena; at some point, this conviction led him to abandon the concept of the ether.
2. He temporarily abandoned Lorentz's theory of electrodynamics, which appears to attach physical significance to absolute motion (i.e., motion with respect to empty space or the ether).
3. He explored the possibility of an alternative electrodynamical theory, which would justify the emission hypothesis about the constancy of the velocity of light relative to its source.
4. Abandoning such attempts, he reexamined Lorentz's theory, at some point focusing his concern on the conflict of his ideas on relative motion with a particular consequence of Lorentz's theory: the independence of velocity of light of the velocity of its source.

5. He recognized that this conflict involves previously tacitly accepted kinematical assumptions about temporal and spatial intervals, leading him to examine the meaning of the concept of the simultaneity of distant events.
6. He defined simultaneity physically, and constructed a new kinematical theory based on the relativity principle and the light principle, thus resolving the apparent conflict between them.

There have been a number of attempts at a detailed reconstruction of Einstein's development of the theory of relativity, attempts which often differ significantly in their conclusions. Such a reconstruction has to take into account other strands in Einstein's work at this time. In particular, by the time he wrote the relativity paper, he no longer regarded Maxwell's electromagnetic theory as universally valid, and had already proposed his light quantum hypothesis (see paper 5). He had also shown that the equipartition theorem, which his work on the foundations of thermodynamics convinced him is valid for the most general classical-mechanical system, combined with Maxwell's theory, leads to an incorrect law for black-body radiation (see section 1 of paper 5). Thus, he already had challenged the unlimited validity of both classical mechanics and of Maxwell's theory.

Einstein later recalled that, uncertain how to proceed in the search for better theories of the structure of matter and radiation, he became convinced that "only the discovery of a universal formal principle could lead . . . to assured results."[7] Such principles play a role analogous in this respect with the role played by the principles of thermodynamics. The theory of relativity is based on just such principles: even though suggested originally by specific mechanical and electromagnetic theories, the principles of relativity and of the constancy of the velocity of light are supported by empirical evidence that is independent of the validity of these theories.

ACCORDING to his sister's memoir, Einstein was anxious about whether his relativity paper would be accepted by the *Annalen der Physik*. After it was accepted, he eagerly anticipated an immediate reaction to its publication, even though he expected it to be critical. He was greatly disappointed when his paper was not even mentioned in the following issues of the *Annalen*. Sometime afterward, she recounts, he received a letter from Planck, requesting explanations of a few obscure points in the work. "After the long period of waiting, this was the first sign that his paper was being read at all. The happiness of the young scholar was all the greater, since acknowledgment of his accomplishment came from one of the greatest contemporary physicists. . . . At that time Planck's interest signified infinitely much for the morale of the young physicist."[8]

Planck and Einstein continued to correspond, and during the fall of 1905 Planck discussed Einstein's paper in the University of Berlin's physics colloquium. During the next few years, Planck wrote several papers developing further consequences of the relativity principle, and interested his assistant, Max Laue, and one of his students, Kurd von Mosengeil, in working on related problems. A few years later, Einstein paid tribute to Planck's role in promoting the theory of relativity: "The attention that this theory so quickly received from colleagues is surely to be ascribed in large part to the resoluteness and warmth with which he intervened for this theory."[9]

Other physicists also started to discuss Einstein's work in 1905 and 1906. Two months after it appeared, Kaufmann cited it in a preliminary report of his recent experiments on the mass of electrons in β-rays. The following year, in a fuller discussion of his results, while noting that the two theories yield the same equations of motion for the electron, he gave the first clear account of the basic theoretical difference between Lorentz's and Einstein's views. Drude, the editor of the *Annalen*, cited Einstein's paper in the second edition of his standard text on optics, as well as in an article on optics in the *Handbuch der Physik*. Wilhelm Röntgen wrote to Einstein asking for copies of his papers on electrodynamics, presumably in connection with a talk Röntgen was to give on the equations of motion of the electron. Sommerfeld, who heard the talk, soon read Einstein's work and was so impressed that he decided to give a colloquium on it. During 1907, Einstein was in correspondence about the theory with Planck, Laue, Wien, and Minkowski. In the same year, he was asked to write a review article on relativity, which appeared in Stark's *Jahrbuch der Radioaktivität* at the end of the year, and a major publishing house inquired about the possibility of a book on his research. A reference by Ehrenfest in 1907 to Einstein's theory as a "closed system" led Einstein to clarify his view of the nature of the theory. By 1908, the theory of relativity, though still controversial and often not clearly distinguished from Lorentz's electron theory, was a major topic of discussion among leading German-speaking physicists.

SINCE the theory of relativity grew out of Einstein's long-standing concern with electrodynamics, and his applications of the theory were primarily in this field, the theory was often looked upon as essentially another version of Lorentz's electron theory. Einstein soon felt the need to make clear the distinction between the kinematical results of the theory, deduced from the two principles of the theory, and the use of such kinematical results in the solution of problems in the optics and electrodynamics of moving bodies, the derivation of the equations of motion of a charged particle—or indeed in any physical theory. He pointed out that the postulates of the theory do not constitute a "closed system" but only a "heuristic principle, which considered by itself alone only contains assertions about rigid bodies, clocks, and light signals." Beyond such assertions, the theory could only establish "relations between otherwise apparently independent laws" of physics.[10]

A few months after first publishing the theory of relativity, Einstein discovered something that particularly intrigued him: the relation between inertial mass and energy. He wrote to Conrad Habicht during the summer of 1905: "One more consequence of the paper on electrodynamics has also occurred to me. The principle of relativity, in conjunction with Maxwell's equations, requires that mass be a direct measure of the energy contained in a body; light carries mass with it. A noticeable decrease of mass should occur in the case of radium. The argument is amusing and seductive; but for all I know the Lord might be laughing over it and leading me around by the nose."

The idea that inertial mass is associated with electromagnetic energy was often discussed before 1905. Around the turn of the century, it was suggested that all mechanical concepts might be derivable from those of electromagnetism. In particular, there were attempts to derive the entire inertial mass of the electron from the energy associated with its electromagnetic field. It was also proved that a radiation-filled container manifests an apparent inertial mass, which (if the mass of the container is neglected) is proportional to the energy of the enclosed radiation.

In paper 4, Einstein argued that, as a consequence of the relativity principle, inertial mass is associated with *all* forms of energy. He was only able to establish this result for a process involving the emission of electromagnetic radiation by a system, but argued that the result is independent of the mechanism by which the system loses energy. In addition, he was only able to show that a change in energy is associated with a change in inertial mass equal to the change in energy divided by c^2. His argument was criticized in 1907 by Planck, who presented his own argument to show that a transfer of heat is associated with a similarly related transfer of inertial mass.

Soon afterward, Stark attributed the discovery of the relation between mass and energy to Planck. Einstein wrote Stark on 17 February 1908: "I was rather disturbed that you do not acknowledge my priority with regard to the connection between inertial mass and energy." After receiving a conciliatory reply from Stark, acknowledging his priority, Einstein replied on 22 February, regretting his original testy reaction: "People, to whom it is granted to contribute something to the progress of science, should not allow pleasure in the fruits of their common work to be clouded by such matters."

Einstein returned to the relation between inertial mass and energy in 1906 and in 1907, giving more general arguments for their complete equivalence, but he did not achieve the complete generality to which he aspired. In his 1909 Salzburg talk, Einstein strongly emphasized that inertial mass is a property of all forms of energy, and therefore electromagnetic radiation must have mass. This conclusion strengthened Einstein's belief in the hypothesis that light quanta manifest particle-like properties.

In 1905, Einstein proposed a number of other experimentally testable consequences of his theory, in particular the equations of motion of the electron. The following year he suggested an experimental test of these equations employing cathode rays.[11]

In this paper, he also mentioned Kaufmann's experimental investigations of the motion of electrons in β-rays. Starting in 1901, Kaufmann had carried out a series of experiments on the deflection of β-rays by electric and magnetic fields. In 1905 he asserted that his recent experiments yielded data for the dependence of mass on velocity that were incompatible with the (identical) predictions of the Lorentz and Einstein theories. Kaufmann's work occasioned considerable discussion. Lorentz was disheartened by the apparent refutation of his theory. Planck subjected the experiment to a careful analysis, and concluded that Kaufmann's work could not be regarded as a definitive refutation of the Lorentz-Einstein predictions. Röntgen, one of the leading German experimentalists, is reported also to have felt that Kaufmann's results were not decisive, because his observations were not that accurate. In a 1907 review, Einstein discussed Kaufmann's results at some length, especially their apparent irreconcilability with the Lorentz-Einstein predictions. Commenting on a figure showing Kaufmann's results and the relativistic predictions, Einstein wrote: "Considering the difficulty of the experiment, one might be inclined to regard the agreement as satisfactory." However, he noted that the deviations were systematic and well outside Kaufmann's error limits. "Whether the systematic deviations are based upon a source of error not yet considered, or on lack of correspondence between the foundations of the theory of relativity and the facts, can only be decided with certainty when more manifold observational data are at hand."[12]

Although Einstein evidently accepted experiment as the ultimate arbiter of the fate of the theory, he was cautious about accepting Kaufmann's results as definitive, perhaps because of his familiarity with Planck's critical analysis of the experiments. What he found even more difficult to accept were alternative equations of motion for the electron that are based on what he regarded as arbitrary dynamical assumptions about the shape of a moving electron. While conceding that Kaufmann's data seemed to favor the theories of Abraham and Bucherer, Einstein concluded: "In my opinion, however, a rather small probability should be ascribed to these theories, since their fundamental assumptions about the mass of a moving electron are not supported by theoretical systems that embrace wider complexes of phenomena."[13]

This cautious attitude toward Kaufmann's results proved justified. During the following years, controversies over the interpretation of the experimental results prevented them from playing a decisive role in contemporary evaluations of the theory of relativity. Bestelmeyer carried out β-ray experiments generally regarded as inconclusive, while Bucherer's results favoring the Lorentz-Einstein equations were seriously questioned. Experiments using cathode rays, reported by several investigators starting in 1910, proved similarly inconclusive. Almost a decade elapsed until data supporting the relativistic predictions (the 1916 results of Guye and Lavanchy) were generally accepted.

EINSTEIN ON THE THEORY OF RELATIVITY

EDITORIAL NOTES

[1]Einstein, "Erinnerungen—Souvenirs," *Schweizerische Hochschulzeitung* 28 (*Sonderheft*) (1955): 146.

[2]John Stuart Mill, *A System of Logic Ratiocinative and Inductive: Being a Connected View of the Principles of Evidence and the Methods of Scientific Investigation*, 8th ed. (London: Longmans, Green, Reader and Dyer, 1872; 1st ed., 1843), vol. 2, pp. 12, 20, 23.

[3]Message by Einstein, prepared for R. S. Shankland to read at a celebration of the centennial of Michelson's birth, 19 December 1952, at Case Institute.

[4]Unpublished manuscript, entitled "Fundamental Ideas and Methods of the Theory of Relativity, Presented as It Developed," copy in Einstein Editorial Archive, Boston University.

[5]Recording, transcribed in Friedrich Herneck, "Zwei Tondokumente Einsteins zur Relativitätstheorie," *Forschungen und Fortschritte* 40 (1966): 134.

[6]See the report of Einstein's talk, given on 14 December 1922, in Jun Ishiwara, *Einstein Kyôzyu-Kôen-roku* (Tokyo: Kabushika Kaisha, 1971), pp. 78–88.

[7]Einstein, *Autobiographical Notes*, Paul Arthur Schilpp, trans. and ed. (La Salle, Ill.: Open Court, 1979), p. 48.

[8]Maja Winteler-Einstein, "Albert Einstein: Beitrag für sein Lebensbild," typescript, pp. 23–24, Einstein Editorial Archive, Boston University.

[9]*Die Naturwissenschaften* 1 (1913): 1079, reprinted in *Collected Papers*, vol. 4, doc. 23, pp. 561–563.

[10]Einstein, "Comments on the Note of Mr. Paul Ehrenfest: 'The Translatory Motion of Deformable Electrons and the Area Law,'" *Collected Papers*, vol. 2, doc. 44, pp. 410-412.

[11]*Annalen der Physik* 21 (1906): 583–586, reprinted in *Collected Papers*, vol. 2, doc. 36, pp. 368–371.

[12]*Jahrbuch der Radioaktivität und Elektronik* 4 (1907): 433–462, citation on pp. 437–439, reprinted in *Collected Papers*, vol. 2, doc. 47, pp. 433–484.

[13]Ibid.

PAPER 3

✦

On the Electrodynamics
of Moving Bodies

IT IS WELL KNOWN that Maxwell's electrodynamics—as usually understood at present—when applied to moving bodies, leads to asymmetries that do not seem to be inherent in the phenomena. Take, for example, the electrodynamic interaction between a magnet and a conductor. The observable phenomenon here depends only on the relative motion of conductor and magnet, whereas the customary view draws a sharp distinction between the two cases, in which either the one or the other of the two bodies is in motion. For if the magnet is in motion and the conductor is at rest, an electric field with a definite energy value results in the vicinity of the magnet that produces a current wherever parts of the conductor are located. But if the magnet is at rest while the conductor is moving, no electric field results in the vicinity of the magnet, but rather an electromotive force in the conductor, to which no energy per se corresponds, but which, assuming an equality of relative motion in the two cases, gives rise to electric currents of the same magnitude and the same course as those produced by the electric forces in the former case.

Examples of this sort, together with the unsuccessful attempts to detect a motion of the earth relative to the "light medium," lead to the conjecture that not only the phenomena of mechanics but also those of electrodynamics have no properties that correspond to the concept of absolute rest. Rather, the same laws of electrodynamics and optics will be valid[1] for all coordinate systems in which the equations of mechanics hold, as has already been shown for quantities of the first order. We shall raise this conjecture (whose content will hereafter be called "the principle of relativity") to the status of a postulate and shall also introduce another postulate, which is only seemingly incompatible with it, namely that light always propagates in empty space with a definite velocity V that is independent of the state of motion of the emitting body. These two postulates suffice for the attainment of a simple and consistent electrodynamics of moving bodies based on Maxwell's theory for bodies at rest. The introduction of a "light ether" will prove to be superfluous, inasmuch as the view to be developed here will not require a "space at absolute rest" endowed with special properties, nor assign a velocity vector to a point of empty space where electromagnetic processes are taking place.

Like all electrodynamics, the theory to be developed here is based on the kinematics of a rigid body, since the assertions of any such theory have to do with the relations among rigid bodies (coordinate systems), clocks, and electromagnetic processes. Insufficient regard for this circumstance is at the root of the difficulties with which the electrodynamics of moving bodies currently has to contend.

A. Kinematic Part

1. Definition of Simultaneity

Consider a coordinate system in which Newton's mechanical equations are valid. To distinguish this system verbally from those to be introduced later, and to make our presentation more precise, we will call it the "rest system."

If a particle is at rest relative to this coordinate system, its position relative to the latter can be determined by means of rigid measuring rods using the methods of Euclidean geometry and expressed in Cartesian coordinates.

If we want to describe the *motion* of a particle, we give the values of its coordinates as functions of time. However, we must keep in mind that a mathematical description of this kind only has physical meaning if we are already clear as to what we understand here by "time." We have to bear in mind that all our judgments involving time are always judgments about *simultaneous events*. If, for example, I say that "the train arrives here at 7 o'clock," that means, more or less, "the pointing of the small hand of my watch to 7 and the arrival of the train are simultaneous events."[1]

It might seem that all difficulties involved in the definition of "time" could be overcome by my substituting "position of the small hand of my watch" for "time." Such a definition is indeed sufficient if a time is to be defined exclusively for the place at which the watch is located; but the definition is no longer satisfactory when series of events occurring at different locations have to be linked temporally, or—what

---

[1] We shall not discuss here the imprecision inherent in the concept of simultaneity of two events taking place at (approximately) the same location, which can be removed only by an abstraction.

amounts to the same thing—when events occurring at places remote from the clock have to be evaluated temporally.

To be sure, we could content ourselves with evaluating the time of events by stationing an observer with a clock at the origin of the coordinates who assigns to an event to be evaluated the corresponding position of the hands of the clock when a light signal from that event reaches him through empty space. However, we know from experience that such a coordination has the drawback of not being independent of the position of the observer with the clock. We reach a far more practical arrangement by the following argument.

If there is a clock at point A in space, then an observer located at A can evaluate the time of events in the immediate vicinity of A by finding the positions of the hands of the clock that are simultaneous with these events. If there is another clock at point B that in all respects resembles the one at A, then the time of events in the immediate vicinity of B can be evaluated by an observer at B. But it is not possible to compare the time of an event at A with one at B without a further stipulation. So far we have defined only an "A-time" and a "B-time," but not a common "time" for A and B. The latter can now be determined by establishing *by definition* that the "time" required for light to travel from A to B is equal to the "time" it requires to travel from B to A. For, suppose a ray of light leaves from A for B at "A-time" t_A, is reflected from B toward A at "B-time" t_B, and arrives back at A at "A-time" t'_A. The two clocks are synchronous by definition if

$$t_B - t_A = t'_A - t_B.$$

We assume that it is possible for this definition of synchronism to be free of contradictions, and to be so for arbitrarily many points, and therefore that the following relations are generally valid:

1. If the clock at B runs synchronously with the clock at A, the clock at A runs synchronously with the clock at B.
2. If the clock at A runs synchronously with the clock at B as well as with the clock at C, then the clocks at B and C also run synchronously relative to each other.

By means of certain (imagined) physical experiments, we have established what is to be understood by synchronous clocks at rest relative to each other and located at different places, and thereby obviously arrived at definitions of "synchronous" and "time." The "time" of an event is the reading obtained simultaneously from a clock at rest that is located at the place of the event, which for all time determinations runs synchronously with a specified clock at rest, and indeed with the specified clock.

Based on experience, we further stipulate that the quantity

$$\frac{2\,\overline{AB}}{t'_A - t_A} = V$$

be a universal constant (the velocity of light in empty space).

It is essential that we have defined time by means of clocks at rest in the rest system; because the time just defined is related to the system at rest, we call it "the time of the rest system."

2. On the Relativity of Lengths and Times

The following considerations are based on the principle of relativity and the principle of the constancy of the velocity of light. We define these two principles as follows:

1. If two coordinate systems are in uniform parallel translational motion relative to each other, the laws according to
which the states of a physical system change do not depend
on which of the two systems these changes are related to.

2. Every light ray moves in the "rest" coordinate system with a
fixed velocity V, independently of whether this ray of light is
emitted by a body at rest or in motion. Hence,

$$\text{velocity} = \frac{\text{light path}}{\text{time interval}},$$

where "time interval" should be understood in the sense of
the definition given in section 1.

Take a rigid rod at rest; let its length, measured by a measuring rod that is also at rest, be l. Now imagine the axis of
the rod placed along the X-axis of the rest coordinate system, and the rod then set into uniform parallel translational
motion (with velocity v) along the X-axis in the direction of
increasing x. We now inquire about the length of the *moving* rod, which we imagine to be ascertained by the following
two operations:

a. The observer moves together with the aforementioned measuring rod and the rigid rod to be measured, and measures
the length of the rod by laying out the measuring rod in the
same way as if the rod to be measured, the observer, and the
measuring rod were all at rest.

b. Using clocks at rest and synchronous in the rest system as
outlined in section 1, the observer determines at which points
of the rest system the beginning and end of the rod to be
measured are located at some given time t. The distance between these two points, measured with the rod used before—
but now at rest—is also a length that we can call the "length
of the rod."

According to the principle of relativity, the length determined by operation (a), which we shall call "the length of the rod in the moving system," must equal the length l of the rod at rest.

The length determined using operation (b), which we shall call "the length of the (moving) rod in the rest system," will be determined on the basis of our two principles, and we shall find that it differs from l.

Current kinematics tacitly assumes that the lengths determined by the above two operations are exactly equal to each other, or, in other words, that at the time t a moving rigid body is totally replaceable, in geometric respects, by the *same* body when it is *at rest* in a particular position.

Further, we imagine the two ends (A and B) of the rod equipped with clocks that are synchronous with the clocks of the rest system, i.e., whose readings always correspond to the "time of the system at rest" at the locations the clocks happen to occupy; hence, these clocks are "synchronous in the rest system."

We further imagine that each clock has an observer co-moving with it, and that these observers apply to the two clocks the criterion for the synchronous rate of two clocks formulated in section 1. Let a ray of light start out from A at time[2] t_A; it is reflected from B at time t_B, and arrives back at A at time t'_A. Taking into account the principle of the constancy of the velocity of light, we find that

$$t_B - t_A = \frac{r_{AB}}{V - v}$$

---

[2] "Time" here means both "time of the system at rest" and "the position of the hands of the moving clock located at the place in question."

and

$$t'_A - t_B = \frac{r_{AB}}{V + v},$$

where r_{AB} denotes the length of the moving rod, measured in the rest system. Observers co-moving with the rod would thus find that the two clocks do not run synchronously, while observers in the system at rest would declare them to be running synchronously.

Thus we see that we cannot ascribe *absolute* meaning to the concept of simultaneity; instead, two events that are simultaneous when observed from some particular coordinate system can no longer be considered simultaneous when observed from a system that is moving relative to that system.

3. Theory of Transformations of Coordinate and Time from the Rest System to a System in Uniform Translational Motion Relative to It

Let there be two coordinate systems in the "rest" space, i.e., two systems of three mutually perpendicular rigid material lines originating from one point. Let the X-axes of the two systems coincide, and their Y- and Z-axes be respectively parallel. Each system shall be supplied with a rigid measuring rod and a number of clocks, and let both measuring rods and all the clocks of the two systems be exactly alike.

Now, put the origin of one of the two systems, say k, in a state of motion with (constant) velocity v in the direction of increasing x of the other system (K), which remains at rest; and let this new velocity be imparted to k's coordinate axes, its corresponding measuring rod, and its clocks. To each time t of the rest system K, there corresponds a definite location of the axes of the moving system. For reasons of symmetry we are justified to assume that the motion of k can be such that at time t ("t" always denotes a time of the rest system) the axes of the moving system are parallel to the axes of the rest system.

We now imagine space to be measured out from both the rest system K using the measuring rod at rest, and from the moving system k using the measuring rod moving along with it, and that coordinates x, y, z and ξ, η, ζ respectively are obtained in this way. Further, by means of the clocks at rest in the rest system, and using light signals as described in section 1, we determine the time t of the rest system for all the points where there are clocks. In a similar manner, by again applying the method of light signals described in section 1, we determine the time τ of the moving system, for all points of this moving system at which there are clocks at rest relative to this system.

To every set of values x, y, z, t which completely determines the place and time of an event in the rest system, there corresponds a set of values ξ, η, ζ, τ that fixes this event relative to the system k, and the problem to be solved now is to find the system of equations that connects these quantities.

First of all, it is clear that these equations must be *linear* because of the properties of homogeneity that we attribute to space and time.

If we put $x' = x - vt$, then it is clear that a point at rest in the system k has a definite, time-independent set of values x', y, z belonging to it. We first determine τ as a function of x', y, z, and t. To this end, we must express in equations that τ is in fact the aggregate of readings of clocks at rest in system k, synchronized according to the rule given in section 1.

Suppose that at time τ_0, a light ray is sent from the origin of the system k along the X-axis to x' and reflected from there toward the origin at time τ_1, arriving there at time τ_2; we then must have

$$\frac{1}{2}(\tau_0 + \tau_2) = \tau_1,$$

or, including the arguments of the function τ and applying the principle of the constancy of the velocity of light in the rest system,

$$\frac{1}{2}\left[\tau(0,0,0,t) + \tau\left(0,0,0,\left\{t + \frac{x'}{V-v} + \frac{x'}{V+v}\right\}\right)\right]$$
$$= \tau\left(x',0,0,t + \frac{x'}{V-v}\right).$$

From this we get, letting x' be infinitesimally small,

$$\frac{1}{2}\left(\frac{1}{V-v} + \frac{1}{V+v}\right)\frac{\partial\tau}{\partial t} = \frac{\partial\tau}{\partial x'} + \frac{1}{V-v}\frac{\partial\tau}{\partial t},$$

or

$$\frac{\partial\tau}{\partial x'} + \frac{v}{V^2 - v^2}\frac{\partial\tau}{\partial t} = 0.$$

It should be noted that, instead of the coordinate origin, we could have chosen any other point as the origin of the light ray, and therefore the equation just derived holds for all values of x', y, z.

Analogous reasoning—applied to the $H^{[2]}$ and Z axes—yields, remembering that light always propagates along these axes with the velocity $\sqrt{V^2 - v^2}$ when observed from the rest system,

$$\frac{\partial\tau}{\partial y} = 0$$

$$\frac{\partial\tau}{\partial z} = 0.$$

These equations yield, since τ is a *linear* function,

$$\tau = a\left(t - \frac{v}{V^2 - v^2}\, x'\right),$$

where a is a function $\varphi(v)$ as yet unknown, and where we assume for brevity that at the origin of k we have $t = 0$ when $\tau = 0$.

Using this result, we can easily determine the quantities ξ, η, ζ by expressing in equations that (as demanded by the principle of the constancy of the velocity of light in conjunction with the principle of relativity) light also propagates with velocity V when measured in the moving system. For a light ray emitted at time $\tau = 0$ in the direction of increasing ξ, we have

$$\xi = V\tau,$$

or

$$\xi = aV\left(t - \frac{v}{V^2 - v^2}\, x'\right).$$

But as measured in the rest system, the light ray propagates with velocity $V - v$ relative to the origin of k, so that

$$\frac{x'}{V - v} = t.$$

Substituting this value of t in the equation for ξ, we obtain

$$\xi = a\, \frac{V^2}{V^2 - v^2}\, x'.$$

Analogously, by considering light rays moving along the two other axes, we get

$$\eta = V\tau = aV\left(t - \frac{v}{V^2 - v^2}\, x'\right),$$

where

$$\frac{y}{\sqrt{V^2 - v^2}} = t; \qquad x' = 0;$$

hence

$$\eta = a \frac{V}{\sqrt{V^2 - v^2}} \, y$$

and

$$\zeta = a \frac{V}{\sqrt{V^2 - v^2}} \, z.$$

If we substitute for x' its value, we obtain

$$\tau = \varphi(v) \beta \left(t - \frac{v}{V^2} \, x \right),$$

$$\xi = \varphi(v) \beta (x - vt),$$

$$\eta = \varphi(v) y,$$

$$\zeta = \varphi(v) z,$$

where

$$\beta = \frac{1}{\sqrt{1 - \left(\dfrac{v}{V} \right)^2}}$$

and φ is an as yet unknown function of v. If no assumptions are made regarding the initial position of the moving system and the zero point of τ, then a constant must be added to the right-hand sides of these equations.

Now we have to prove that, measured in the moving system, every light ray propagates with the velocity V, if it does so, as we have assumed, in the rest system; for we have not yet proved that the principle of the constancy of the velocity of light is compatible with the relativity principle.

Suppose that at time $t = \tau = 0$ a spherical wave is emitted from the coordinate origin, which at that time is common to both systems, and that this wave propagates in the system K with the velocity V. Hence, if (x, y, z) is a point reached by this wave, we have

$$x^2 + y^2 + z^2 = V^2 t^2.$$

We transform this equation using our transformation equations and, after a simple calculation, obtain

$$\xi^2 + \eta^2 + \zeta^2 = V^2 \tau^2.$$

Thus, our wave is also a spherical wave with propagation velocity V when it is observed in the moving system. This proves that our two fundamental principles are compatible.[3]

The transformation equations we have derived also contain an unknown function φ of v, which we now wish to determine.

To this end we introduce a third coordinate system K', which, relative to the system k, is in parallel-translational motion, parallel to the axis Ξ,[4] such that its origin moves along the Ξ-axis with velocity $-v$. Let all three coordinate origins coincide at time $t = 0$, and let the time t' of system K' equal zero at $t = x = y = z = 0$. We denote the coordinates measured in the system K' by x', y', z' and, by twofold application of our transformation equations, we get

$$t' = \varphi(-v)\beta(-v)\left\{\tau + \frac{v}{V^2}\xi\right\} = \varphi(v)\varphi(-v)t,$$

$$x' = \varphi(-v)\beta(-v)\{\xi + v\tau\} \quad = \varphi(v)\varphi(-v)x,$$

$$y' = \varphi(-v)\eta \qquad\qquad = \varphi(v)\varphi(-v)y,$$

$$z' = \varphi(-v)\zeta \qquad\qquad = \varphi(v)\varphi(-v)z.$$

Since the relations between x', y', z' and x, y, z do not contain the time t, the systems K and K' are at rest relative to each other, and it is clear that the transformation from K to K' must be the identity transformation. Hence,

$$\varphi(v)\varphi(-v) = 1.$$

Let us now explore the meaning of $\varphi(v)$. We shall focus on that portion of the H-axis of the system k that lies between $\xi = 0$, $\eta = 0$, $\zeta = 0$, and $\xi = 0$, $\eta = l$, $\zeta = 0$. This portion of the H-axis is a rod that, relative to the system K, moves perpendicular to its axis with a velocity v and its ends have coordinates in K:

$$x_1 = vt, \quad y_1 = \frac{l}{\varphi(v)}, \quad z_1 = 0$$

and

$$x_2 = vt, \quad y_2 = 0, \quad z_2 = 0.$$

The length of the rod, measured in K, is thus $l/\varphi(v)$; this gives us the meaning of the function φ. For reasons of symmetry, it is now evident that the length of a rod measured in the rest system and moving perpendicular to its axis can depend only on its velocity and not on the direction and sense of its motion. Thus, the length of the moving rod measured in the rest system does not change if v is replaced by $-v$. From this we conclude:

$$\frac{l}{\varphi(v)} = \frac{l}{\varphi(-v)},$$

or

$$\varphi(v) = \varphi(-v).$$

From this relation and the one found earlier it follows that $\varphi(v) = 1$, so that the transformation equations obtained become

$$\tau = \beta\left(t - \frac{v}{V^2}\,x\right),$$

$$\xi = \beta(x - vt),$$

$$\eta = y,$$

$$\zeta = z,$$

where

$$\beta = \frac{1}{\sqrt{1 - \left(\dfrac{v}{V}\right)^2}}.$$

4. The Physical Meaning of the Equations Obtained as Concerns Moving Rigid Bodies and Moving Clocks

We consider a rigid sphere[3] of radius R that is at rest relative to the moving system k and whose center lies at the origin of k. The equation of the surface of this sphere, which moves with velocity v relative to the system k, is

$$\xi^2 + \eta^2 + \zeta^2 = R^2.$$

Expressed in terms of x, y, z, the equation of this surface at time $t = 0$ is

$$\frac{x^2}{\left(\sqrt{1 - \left(\dfrac{v}{V}\right)^2}\right)^2} + y^2 + z^2 = R^2.$$

A rigid body that has a spherical shape when measured at rest has, when in motion—considered from the rest

---

[3] I.e., a body that has a spherical shape when examined at rest.

system—the shape of an ellipsoid of revolution with axes

$$R\sqrt{1-\left(\frac{v}{V}\right)^2},\ R,\ R.$$

Thus, while the Y and Z dimensions of the sphere (and hence also of every rigid body, whatever its shape) do not appear to be altered by motion, the X dimension appears to be contracted in the ratio $1 : \sqrt{1-(v/V)^2}$, thus the greater the value of v, the greater the contraction. For $v = V$, all moving objects—considered from the "rest" system—shrink into plane structures. For superluminal velocities our considerations become meaningless; as we shall see from later considerations, in our theory the velocity of light physically plays the role of infinitely great velocities.

It is clear that the same results apply for bodies at rest in the "rest" system when considered from a uniformly moving system.

We further imagine one of the clocks that is able to indicate time t when at rest relative to the rest system and time τ when at rest relative to the moving system to be placed at the origin of k and set such that it indicates the time τ. What is the rate of this clock when considered from the rest system?

The quantities x, t, and τ that refer to the position of this clock obviously satisfy the equations

$$\tau = \frac{1}{\sqrt{1-\left(\frac{v}{V}\right)^2}}\left(t - \frac{v}{V^2}x\right)$$

and

$$x = vt.$$

We thus have

$$\tau = t\sqrt{1 - \left(\frac{v}{V}\right)^2} = t - \left(1 - \sqrt{1 - \left(\frac{v}{V}\right)^2}\right)t,$$

from which it follows that the reading of the clock considered from the rest system lags behind each second by $(1 - \sqrt{1 - (v/V)^2})$ sec or, up to quantities of the fourth and higher order, by $\frac{1}{2}(v/V)^2$ sec.

This yields the following peculiar consequence: If at the points A and B of K there are clocks at rest that, considered from the rest system, are running synchronously, and if the clock at A is transported to B along the connecting line with velocity v, then upon arrival of this clock at B the two clocks will no longer be running synchronously; instead, the clock that has been transported from A to B will lag $\frac{1}{2}tv^2/V^2$ sec (up to quantities of the fourth and higher orders) behind the clock that has been in B from the outset, where t is the time needed by the clock to travel from A to B.

We see at once that this result holds even when the clock moves from A to B along any arbitrary polygonal line, and even when the points A and B coincide.[5]

If we assume that the result proved for a polygonal line holds also for a continuously curved line, then we arrive at the following result: If there are two synchronously running clocks at A, and one of them is moved along a closed curve with constant velocity until it has returned to A, which takes, say, t sec, then, on its arrival at A, this clock will lag $\frac{1}{2}t(v/V)^2$ sec behind the clock that has not been moved. From this we conclude that a balance-wheel clock[6] located at the Earth's equator must, under otherwise identical conditions, run more slowly by a very small amount than an absolutely identical clock located at one of the Earth's poles.

5. *The Addition Theorem for Velocities*

In the system k moving with velocity v along the X-axis of the system K, let a point move in accordance with the equations

$$\xi = w_\xi \tau,$$

$$\eta = w_\eta \tau,$$

$$\zeta = 0,$$

where w_ξ and w_η denote constants.

We seek the motion of the point relative to the system K. Introducing the quantities x, y, z, t into the equations of motion of the point by means of the transformation equations derived in section 3, we obtain

$$x = \frac{w_\xi + v}{1 + \dfrac{vw_\xi}{V^2}}\, t,$$

$$y = \frac{\sqrt{1 - \left(\dfrac{v}{V}\right)^2}}{1 + \dfrac{vw_\xi}{V^2}}\, w_\eta t,$$

$$z = 0.$$

Thus, according to our theory, the vector addition for velocities holds only to first approximation. Let

$$U^2 = \left(\frac{dx}{dt}\right)^2 + \left(\frac{dy}{dt}\right)^2,$$

$$w^2 = w_\xi^2 + w_\eta^2$$

and

$$\alpha = \arctan \frac{w_y}{w_x}; \text{[7]}$$

α is then to be considered as the angle between the velocities v and w. After a simple calculation we obtain

$$U = \frac{\sqrt{(v^2 + w^2 + 2vw\cos\alpha) - \left(\dfrac{vw\sin\alpha}{V}\right)^2}}{1 + \dfrac{vw\cos\alpha}{V^2}}.$$

It is worth noting that v and w enter into the expression for the resultant velocity in a symmetrical manner. If w also has the direction of the X-axis (Ξ-axis), we get

$$U = \frac{v + w}{1 + \dfrac{vw}{V^2}}.$$

It follows from this equation that the composition of two velocities that are smaller than V always results in a velocity that is smaller than V. For if we set $v = V - \kappa$, and $w = V - \lambda$, where κ and λ are positive and smaller than V, then

$$U = V\,\frac{2V - \kappa - \lambda}{2V - \kappa - \lambda + \dfrac{\kappa\lambda}{V}} < V.$$

It also follows that the velocity of light V cannot be changed by compounding it with a "subluminal velocity." For this case we get

$$U = \frac{V + w}{1 + \dfrac{w}{V}} = V.$$

In the case where v and w have the same direction, the formula for U could also have been obtained by compounding two transformations according to section 3. If in addition to the systems K and k, occurring in section 3, we introduce a third coordinate system k', which moves parallel to k and whose origin moves with velocity w along the Ξ-axis, we obtain equations between the quantities x, y, z, t and the

141

corresponding quantities of k' that differ from those found in section 3 only insofar as "v''" is replaced by the quantity

$$\frac{v+w}{1+\dfrac{vw}{V^2}};$$

from this we see that such parallel transformations form a group—as indeed they must.

We have now derived the required laws of the kinematics corresponding to our two principles, and proceed to their application to electrodynamics.

B. ELECTRODYNAMIC PART

6. Transformation of the Maxwell-Hertz Equations for Empty Space. On the Nature of the Electromotive Forces Arising Due to Motion in a Magnetic Field

Let the Maxwell-Hertz equations for empty space be valid for the rest system K, so that we have

$$\frac{1}{V}\frac{\partial X}{\partial t} = \frac{\partial N}{\partial y} - \frac{\partial M}{\partial z}, \qquad \frac{1}{V}\frac{\partial L}{\partial t} = \frac{\partial Y}{\partial z} - \frac{\partial Z}{\partial y},$$

$$\frac{1}{V}\frac{\partial Y}{\partial t} = \frac{\partial L}{\partial z} - \frac{\partial N}{\partial x}, \qquad \frac{1}{V}\frac{\partial M}{\partial t} = \frac{\partial Z}{\partial x} - \frac{\partial X}{\partial z},$$

$$\frac{1}{V}\frac{\partial Z}{\partial t} = \frac{\partial M}{\partial x} - \frac{\partial L}{\partial y}, \qquad \frac{1}{V}\frac{\partial N}{\partial t} = \frac{\partial X}{\partial y} - \frac{\partial Y}{\partial x},$$

where (X, Y, Z) denotes the electric force vector and (L, M, N) the magnetic force vector.

If we apply the transformations derived in section 3 to these equations, in order to relate the electromagnetic processes to the coordinate system moving with velocity v introduced there, we obtain the following equations:

$$\frac{1}{V}\frac{\partial X}{\partial \tau} = \frac{\partial \beta \left(N - \frac{v}{V}Y\right)}{\partial \eta} - \frac{\partial \beta \left(M + \frac{v}{V}Z\right)}{\partial \zeta},$$

$$\frac{1}{V}\frac{\partial \beta \left(Y - \frac{v}{V}N\right)}{\partial \tau} = \frac{\partial L}{\partial \zeta} - \frac{\partial \beta \left(N - \frac{v}{V}Y\right)}{\partial \xi},$$

$$\frac{1}{V}\frac{\partial \beta \left(Z + \frac{v}{V}M\right)}{\partial \tau} = \frac{\partial \beta \left(M + \frac{v}{V}Z\right)}{\partial \xi} - \frac{\partial L}{\partial \eta},$$

$$\frac{1}{V}\frac{\partial L}{\partial \tau} = \frac{\partial \beta \left(Y - \frac{v}{V}N\right)}{\partial \zeta} - \frac{\partial \beta \left(Z + \frac{v}{V}M\right)}{\partial \eta},$$

$$\frac{1}{V}\frac{\partial \beta \left(M + \frac{v}{V}Z\right)}{\partial \tau} = \frac{\partial \beta \left(Z + \frac{v}{V}M\right)}{\partial \xi} - \frac{\partial X}{\partial \zeta},$$

$$\frac{1}{V}\frac{\partial \beta \left(N - \frac{v}{V}Y\right)}{\partial \tau} = \frac{\partial X}{\partial \eta} - \frac{\partial \beta \left(Y - \frac{v}{V}N\right)}{\partial \xi},$$

where

$$\beta = \frac{1}{\sqrt{1 - \left(\frac{v}{V}\right)^2}}.$$

The relativity principle requires that the Maxwell-Hertz equations for empty space also be valid in the system k if they are valid in the system K, i.e, that the electric and magnetic force vectors—(X', Y', Z') and (L', M', N')—of the moving system k, which are defined in this system by their ponderomotive effects on electric and magnetic charges, respectively, satisfy the equations

$$\frac{1}{V}\frac{\partial X'}{\partial \tau} = \frac{\partial N'}{\partial \eta} - \frac{\partial M'}{\partial \zeta}, \qquad \frac{1}{V}\frac{\partial L'}{\partial \tau} = \frac{\partial Y'}{\partial \zeta} - \frac{\partial Z'}{\partial \eta},$$

$$\frac{1}{V}\frac{\partial Y'}{\partial \tau} = \frac{\partial L'}{\partial \zeta} - \frac{\partial N'}{\partial \xi}, \qquad \frac{1}{V}\frac{\partial M'}{\partial \tau} = \frac{\partial Z'}{\partial \xi} - \frac{\partial X'}{\partial \zeta},$$

$$\frac{1}{V}\frac{\partial Z'}{\partial \tau} = \frac{\partial M'}{\partial \xi} - \frac{\partial L'}{\partial \eta}, \qquad \frac{1}{V}\frac{\partial N'}{\partial \tau} = \frac{\partial X'}{\partial \eta} - \frac{\partial Y'}{\partial \xi}.$$

Obviously, the two systems of equations found for the system k must express exactly the same thing, since both systems of equations are equivalent to the Maxwell-Hertz equations for the system K. Further, since the equations for the two systems are in agreement apart from the symbols representing the vectors, it follows that the functions occurring in the systems of equations at corresponding places must agree up to a factor $\psi(v)$, common to all functions of one of the systems of equations and independent of ξ, η, ζ, and τ, but possibly depending on v. Thus we have the relations:

$$X' = \psi(v)X, \qquad L' = \psi(v)L,$$

$$Y' = \psi(v)\beta\left(Y - \frac{v}{V}N\right), \qquad M' = \psi(v)\beta\left(M + \frac{v}{V}Z\right),$$

$$Z' = \psi(v)\beta\left(Z + \frac{v}{V}M\right), \qquad N' = \psi(v)\beta\left(N - \frac{v}{V}Y\right).$$

If we now invert this system of equations, first by solving the equations just obtained, and second by applying to the equations the inverse transformation (from k to K), which is characterized by the velocity $-v$, we get, taking into account that both systems of equations so obtained must be identical,

$$\varphi(v) \cdot \varphi(-v) = 1.$$

Further, it follows for reasons of symmetry[4] that

$$\varphi(v) = \varphi(-v);$$

---

[4] If, e.g., $X = Y = Z = L = M = 0$ and $N \neq 0$, then it is clear for reasons of symmetry that if v changes its sign without changing its numerical value, then Y' must also change its sign without changing its numerical value.

thus

$$\varphi(v) = 1,$$

and our equations take the form

$$X' = X, \qquad L' = L,$$

$$Y' = \beta\left(Y - \frac{v}{V}N\right), \qquad M' = \beta\left(M + \frac{v}{V}Z\right),$$

$$Z' = \beta\left(Z + \frac{v}{V}M\right), \qquad N' = \beta\left(N - \frac{v}{V}Y\right).$$

By way of interpreting these equations, we note the following remarks: Imagine a pointlike electric charge, whose magnitude measured in the rest system is "unit," i.e., which, when at rest in the rest system exerts a force of 1 dyne on an equal charge at a distance of 1 cm. According to the principle of relativity this electric charge is also of "unit" magnitude if measured in the moving system. If this electric charge is at rest relative to the rest system, then by definition the vector (X, Y, Z) equals the force acting on it. If, on the other hand, this acting charge is at rest relative to the moving system (at least at the relevant instant), then the force acting on it measured in the moving system is equal to the vector (X', Y', Z'). Hence, the first three of the above equations can be expressed in words in the following two ways:

1. If a unit point electric charge moves in an electromagnetic field, there acts upon it, in addition to the electric force, an "electromotive force" that, neglecting terms multiplied by the second and higher powers of v/V, is equal to the vector product of the velocity of the charge and the magnetic force, divided by the velocity of light. (Old mode of expression.)

2. If a unit point electric charge moves in an electromagnetic field, the force acting on it equals the electric force at the

location of the unit charge that is obtained by transforming the field to a coordinate system at rest relative to the unit charge. (New mode of expression.)

Analogous remarks hold for "magnetomotive forces."[8] We can see that in the theory developed here, the electromotive force only plays the role of an auxiliary concept, which owes its introduction to the circumstance that the electric and magnetic forces do not have an existence independent of the state of motion of the coordinate system.

It is further clear that the asymmetry in the treatment of currents produced by the relative motion of a magnet and a conductor, mentioned in the introduction, disappears. Moreover, questions about the "site" of electrodynamic electromotive forces (unipolar machines) become pointless.

7. Theory of Doppler's Principle and of Aberration

In the system K and very far from the coordinate origin, let there be a source of electrodynamic waves that, in a part of space containing the coordinate origin, are represented with sufficient accuracy by the equations

$$X = X_0 \sin \Phi, \quad L = L_0 \sin \Phi,$$

$$Y = Y_0 \sin \Phi, \quad M = M_0 \sin \Phi, \quad \Phi = \omega\left(t - \frac{ax + by + cz}{V}\right).$$

$$Z = Z_0 \sin \Phi, \quad N = N_0 \sin \Phi,$$

Here (X_0, Y_0, Z_0) and (L_0, M_0, N_0) are the vectors determining the amplitude of the wave train, and a, b, c are the direction cosines of the norm to the waves.

We want to know the character of these waves when investigated by an observer at rest in the moving system k.

146

Applying the transformation equations for electric and magnetic forces found in section 6 and those for coordinates and time found in section 3, we immediately obtain:

$$X' = X_0 \sin \Phi', \qquad L' = L_0 \sin \Phi',$$

$$Y' = \beta\left(Y_0 - \frac{v}{V}N_0\right)\sin \Phi', \qquad M' = \beta\left(M_0 + \frac{v}{V}Z_0\right)\sin \Phi',$$

$$Z' = \beta\left(Z_0 + \frac{v}{V}M_0\right)\sin \Phi', \qquad N' = \beta\left(N_0 - \frac{v}{V}Y_0\right)\sin \Phi',$$

$$\Phi' = \omega'\left(\tau - \frac{a'\xi + b'\eta + c'\zeta}{V}\right),$$

where we have put

$$\omega' = \omega\beta\left(1 + a\frac{v}{V}\right),$$

$$a' = \frac{a - \dfrac{v}{V}}{1 - a\dfrac{v}{V}},$$

$$b' = \frac{b}{\beta\left(1 - a\dfrac{v}{V}\right)},$$

$$c' = \frac{c}{\beta\left(1 - a\dfrac{v}{V}\right)}.$$

From the equation for ω' it follows that if an observer moves with velocity v relative to an infinitely distant source of light of frequency ν, in such a way that the connecting line "light source–observer" forms an angle φ with the observer's velocity, where this velocity is relative to a coordinate system at rest relative to the light source, then ν', the frequency of the light perceived by the observer, is given by the equation

$$\nu' = \nu \frac{1 - \cos \varphi \dfrac{v}{V}}{\sqrt{1 - \left(\dfrac{v}{V}\right)^2}}.$$

This is Doppler's principle for arbitrary velocities. For $\varphi = 0$ the equation takes the simple form

$$\nu' = \nu \sqrt{\frac{1 - \dfrac{v}{V}}{1 + \dfrac{v}{V}}}.$$

We see that, contrary to the usual conception, when $v = -\infty$, then $\nu = \infty$.[9]

If φ' denotes the angle between the wave normal (the direction of the ray) in the moving system and the connecting line "light source–observer,"[10] the equation for α'[11] takes the form

$$\cos \varphi' = \frac{\cos \varphi - \dfrac{v}{V}}{1 - \dfrac{v}{V} \cos \varphi}.$$

This equation expresses the law of aberration in its most general form. If $\varphi = \pi/2$, the equation takes the simple form

$$\cos \varphi' = -\frac{v}{V}.$$

We still need to find the amplitude of the waves as it appears in the moving system. If A and A' denote the amplitude of electric or magnetic force in the rest system and moving system respectively, we get

$$A'^2 = A^2 \frac{\left(1 - \dfrac{v}{V} \cos \varphi\right)^2}{1 - \left(\dfrac{v}{V}\right)^2},$$

which for $\varphi = 0$ takes the simpler form:

$$A'^2 = A^2 \frac{1 - \dfrac{v}{V}}{1 + \dfrac{v}{V}}.$$

It follows from these results that to an observer approaching a light source with velocity V, this source would have to appear infinitely intense.

8. Transformation of the Energy of Light Rays.
Theory of Radiation Pressure
Exerted on Perfect Mirrors

Since $A^2/8\pi$ equals the energy of light per unit volume, according to the principle of relativity we have to consider $A'^2/8\pi$ as the light energy in the moving system. Hence A'^2/A^2 would be the ratio of the energy of a given light complex "measured in motion" and "measured at rest" if the volume of a light complex were the same measured in K and in k. However, this is not the case. If a, b, c are the direction cosines of the wave normal of the light in the rest system, then the surface elements of the spherical surface

$$(x - Vat)^2 + (y - Vbt)^2 + (z - Vct)^2 = R^2$$

moving with the velocity of light are not traversed by any energy; we may therefore say that this surface permanently encloses the same light complex. We investigate the quantity of energy enclosed by this surface considered from the system k, i.e., the energy of the light complex relative to the system k.

Considered in the moving system, the spherical surface is an ellipsoidal surface whose equation at time $\tau = 0$ is

$$\left(\beta\xi - a\beta\frac{v}{V}\xi\right)^2 + \left(\eta - b\beta\frac{v}{V}\xi\right)^2 + \left(\zeta - c\beta\frac{v}{V\xi}\right)^2 = R^2.$$

If S denotes the volume of the sphere and S' that of the ellipsoid, then a simple calculation shows that

$$\frac{S'}{S} = \frac{\sqrt{1 - \left(\frac{v}{V}\right)^2}}{1 - \frac{v}{V}\cos\varphi}.$$

If we call the energy of the light enclosed by this surface E when measured in the rest system and E' when measured in the moving system, we obtain

$$\frac{E'}{E} = \frac{\frac{A'^2}{8\pi}S'}{\frac{A^2}{8\pi}S} = \frac{1 - \frac{v}{V}\cos\varphi}{\sqrt{1 - \left(\frac{v}{V}\right)^2}},$$

which, for $\varphi = 0$, simplifies to

$$\frac{E'}{E} = \sqrt{\frac{1 - \frac{v}{V}}{1 + \frac{v}{V}}}.$$

It is noteworthy that the energy and the frequency of a light complex vary with the observer's state of motion according to the same law.

Let the coordinate plane $\xi = 0$ be a completely reflecting surface, from which the plane waves considered in section 7 are reflected. We investigate the pressure of light exerted on the reflecting surface, and the direction, frequency, and intensity of the light after reflection.

Let the incident light be defined by the quantities A, cos φ, and ν (relative to system K). Considered from k, the corresponding quantities are

$$A' = A \frac{1 - \frac{v}{V} \cos \varphi}{\sqrt{1 - \left(\frac{v}{V}\right)^2}},$$

$$\cos \varphi' = \frac{\cos \varphi - \frac{v}{V}}{1 - \frac{v}{V} \cos \varphi},$$

$$\nu' = \nu \frac{1 - \frac{v}{V} \cos \varphi}{\sqrt{1 - \left(\frac{v}{V}\right)^2}}.$$

Referring the process to the system k, we get for the reflected light

$$A'' = A',$$

$$\cos \varphi'' = -\cos \varphi',$$

$$\nu'' = \nu'.$$

Finally, by transforming back to the rest system K, we get for the reflected light

$$A''' = A'' \frac{1 + \frac{v}{V} \cos \varphi''}{\sqrt{1 - \left(\frac{v}{V}\right)^2}} = A \frac{1 - 2\frac{v}{V} \cos \varphi + \left(\frac{v}{V}\right)^2}{1 - \left(\frac{v}{V}\right)^2},$$

$$\cos \varphi''' = \frac{\cos \varphi'' + \frac{v}{V}}{1 + \frac{v}{V} \cos \varphi''} = -\frac{\left(1 + \left(\frac{v}{V}\right)^2\right) \cos \varphi - 2\frac{v}{V}}{1 - 2\frac{v}{V} \cos \varphi + \left(\frac{v}{V}\right)^2},$$

$$v''' = v'' \frac{1 + \dfrac{v}{V}\cos\varphi''}{\sqrt{1 - \left(\dfrac{v}{V}\right)^2}} = v \frac{1 - 2\dfrac{v}{V}\cos\varphi + \left(\dfrac{v}{V}\right)^2}{\left(1 - \dfrac{v}{V}\right)^2}. \quad [12]$$

The energy (measured in the rest system) that strikes a unit surface of the mirror per unit time is obviously $A^2/8\pi(V\cos\varphi - v)$. The energy leaving a unit surface of the mirror per unit time is $A'''^2/8\pi(-V\cos\varphi''' + v)$. According to the principle of energy conservation, the difference of these two expressions is the work done by light pressure per unit time. Equating this work to $P \cdot v$, where P is the pressure of light, we obtain

$$P = 2 \frac{A^2}{8\pi} \frac{\left(\cos\varphi - \dfrac{v}{V}\right)^2}{1 - \left(\dfrac{v}{V}\right)^2}.$$

To first approximation, in agreement with experiment and with other theories, we get

$$P = 2 \frac{A^2}{8\pi} \cos^2 \varphi.$$

All problems in the optics of moving bodies can be solved by the method employed here. The essential point is that the electric and magnetic fields of light that is influenced by a moving body are transformed to a coordinate system that is at rest relative to that body. By this means, all problems in the optics of moving bodies are reduced to a series of problems in the optics of bodies at rest.

9. Transformation of the Maxwell-Hertz Equations When Convection Currents Are Taken into Account

We start from the equations

$$\frac{1}{V}\left\{u_x\rho + \frac{\partial X}{\partial t}\right\} = \frac{\partial N}{\partial y} - \frac{\partial M}{\partial z}, \qquad \frac{1}{V}\frac{\partial L}{\partial t} = \frac{\partial Y}{\partial z} - \frac{\partial Z}{\partial y},$$

$$\frac{1}{V}\left\{u_y\rho + \frac{\partial Y}{\partial t}\right\} = \frac{\partial L}{\partial z} - \frac{\partial N}{\partial x}, \qquad \frac{1}{V}\frac{\partial M}{\partial t} = \frac{\partial Z}{\partial x} - \frac{\partial X}{\partial z},$$

$$\frac{1}{V}\left\{u_z\rho + \frac{\partial Z}{\partial t}\right\} = \frac{\partial M}{\partial x} - \frac{\partial L}{\partial y}, \qquad \frac{1}{V}\frac{\partial N}{\partial t} = \frac{\partial X}{\partial y} - \frac{\partial Y}{\partial x},$$

where

$$\rho = \frac{\partial X}{\partial x} + \frac{\partial Y}{\partial y} + \frac{\partial Z}{\partial z}$$

denotes 4π times the charge density, and (u_x, u_y, u_z) the velocity vector of the charge. If the electric charges are conceived as permanently bound to small, rigid bodies (ions, electrons), then these equations constitute the electromagnetic foundation of Lorentz's electrodynamics and optics of moving bodies.

If, using the transformation equations presented in sections 3 and 6, we transform these equations, assumed to be valid in the system K, to the system k, we get the equations

$$\frac{1}{V}\left\{u_\xi\rho' + \frac{\partial X'}{\partial \tau}\right\} = \frac{\partial N'}{\partial \eta} - \frac{\partial M'}{\partial \zeta}, \qquad \frac{1}{V}\frac{\partial L'}{\partial \tau} = \frac{\partial Y'}{\partial \zeta} - \frac{\partial Z'}{\partial \eta},$$

$$\frac{1}{V}\left\{u_\eta\rho' + \frac{\partial Y'}{\partial \tau}\right\} = \frac{\partial L'}{\partial \zeta} - \frac{\partial N'}{\partial \xi}, \qquad \frac{1}{V}\frac{\partial M'}{\partial \tau} = \frac{\partial Z'}{\partial \xi} - \frac{\partial X'}{\partial \zeta},$$

$$\frac{1}{V}\left\{u_\zeta\rho' + \frac{\partial Z'}{\partial \tau}\right\} = \frac{\partial M'}{\partial \xi} - \frac{\partial L'}{\partial \eta}, \qquad \frac{1}{V}\frac{\partial N'}{\partial \tau} = \frac{\partial X'}{\partial \eta} - \frac{\partial Y'}{\partial \xi},$$

where

$$\frac{u_x - v}{1 - \dfrac{u_x v}{V^2}} = u_\xi,$$

$$\frac{u_y}{\beta\left(1 - \dfrac{u_x v}{V^2}\right)} = u_\eta,$$

$$\frac{u_z}{\beta\left(1 - \dfrac{u_x v}{V^2}\right)} = u_\zeta,$$

and

$$\rho' = \frac{\partial X'}{\partial \xi} + \frac{\partial Y'}{\partial \eta} + \frac{\partial Z'}{\partial \zeta} = \beta\left(1 - \frac{v u_x}{V^2}\right)\rho.$$

Since—as follows from the velocity addition theorem (sec. 5)—the vector (u_ξ, u_η, u_ζ) is actually the velocity of the electric charges measured in the system k, we have thus shown that, on the basis of our kinematic principles, the electrodynamic foundation of Lorentz's theory of the electrodynamics of moving bodies agrees with the principle of relativity.

Let me also briefly add that the following important proposition can easily be deduced from the equations we have derived: If an electrically charged body moves arbitrarily in space without altering its charge when observed from a coordinate system moving with the body, then its charge also remains constant when observed from the "rest" system K.

10. Dynamics of the (Slowly Accelerated) Electron

In an electromagnetic field let there move an electrically charged particle with charge ϵ (called an "electron" in what follows); we assume only the following about its law of motion:

If the electron is at rest at a particular moment, its motion during the next instant of time will occur according to the equations

$$\mu\frac{d^2x}{dt^2} = \epsilon X,$$

$$\mu\frac{d^2y}{dt^2} = \epsilon Y,$$

$$\mu\frac{d^2z}{dt^2} = \epsilon Z,$$

where x, y, z denote the coordinates of the electron and μ its mass as long as the electron moves slowly.

Further, let the electron's velocity at a certain moment be v. We investigate the law of motion of the electron during the immediately succeeding instant of time.

Without loss of generality, we may and shall assume that the electron is at the coordinate origin and moves with velocity v along the X-axis of the system K at the moment with which we are concerned. It is then obvious that at the given moment ($t = 0$), the electron is at rest relative to a coordinate system k moving with constant velocity v parallel to the X-axis.

From the above assumption combined with the relativity principle, it is clear that, considered from the system k, the electron will move during the immediately ensuing period of time (for small values of t) according to the equations

$$\mu\frac{d^2\xi}{d\tau^2} = \epsilon X',$$

$$\mu\frac{d^2\eta}{d\tau^2} = \epsilon Y',$$

$$\mu\frac{d^2\zeta}{d\tau^2} = \epsilon Z',$$

where the symbols ξ, η, ζ, τ, X', Y', Z' all refer to the system k. If we also stipulate that, for $t = x = y = z = 0$, $\tau = \xi = \eta = \zeta = 0$ shall also hold, then the transformation equations of sections 3 and 6 are applicable, so that we get

$$\tau = \beta\left(t - \frac{v}{V^2}x\right),$$

$$\xi = \beta(x - vt), \qquad X' = x,$$

$$\eta = y, \qquad Y' = \beta\left(Y - \frac{v}{V}N\right),$$

$$\zeta = z, \qquad Z' = \beta\left(Z + \frac{v}{V}M\right).$$

With the help of these equations we transform the above equations of motion from the system k to the system K, obtaining

$$\frac{d^2x}{dt^2} = \frac{\epsilon}{\mu}\frac{1}{\beta^3}X,$$

$$\frac{d^2y}{dt^2} = \frac{\epsilon}{\mu}\frac{1}{\beta}\left(Y - \frac{v}{V}N\right), \qquad \text{(A)}$$

$$\frac{d^2z}{dt^2} = \frac{\epsilon}{\mu}\frac{1}{\beta}\left(Z + \frac{v}{V}M\right).$$

Following the usual approach, we now investigate the "longitudinal" and "transverse" mass of the moving electron. We write equations (A) in the form

$$\mu\beta^3\frac{d^2x}{dt^2} = \epsilon X = \epsilon X',$$

$$\mu\beta^2\frac{d^2y}{dt^2} = \epsilon\beta\left(Y - \frac{v}{V}N\right) = \epsilon Y',$$

$$\mu\beta^2\frac{d^2z}{dt^2} = \epsilon\beta\left(Z + \frac{v}{V}M\right) = \epsilon Z',$$

and note first that $\epsilon X'$, $\epsilon Y'$, $\epsilon Z'$ are the components of the ponderomotive force acting on the electron, as considered in a moving system that, at this instant, is moving with the same velocity as the electron. (This force could be measured, for example, by a spring balance at rest in the latter system.) If we simply call[13] this force "the force acting on the electron," and preserve the equation

$$\text{Mass} \times \text{Acceleration} = \text{Force},$$

stipulating, in addition, that accelerations be measured in the rest system K, then the above equations lead to the definition:

$$\text{Longitudinal mass} = \frac{\mu}{\left(\sqrt{1 - \left(\frac{v}{V}\right)^2}\right)^3},$$

$$\text{Transverse mass} = \frac{\mu}{1 - \left(\frac{v}{V}\right)^2}.$$

Of course, with a different definition of force and acceleration we would obtain different values for these masses; this shows that we must proceed very cautiously when comparing various theories of electron motion.

It should be noted that these results about mass are also valid for ponderable material points, because a ponderable material point can be made into an electron (in our sense of the word) by adding to it an *arbitrarily small* electric charge.

We now determine the kinetic energy of an electron. If an electron starts from the origin of the system K with an initial velocity 0 and continues to move along the X-axis under the influence of an electrostatic force X, it is clear that the energy it takes from the electrostatic field has the value

$\int \epsilon X dx$. Since the electron is supposed to accelerate slowly, and consequently cannot emit any energy in the form of radiation, the energy taken from the electrostatic field must be equated to the kinetic energy W of the electron. Bearing in mind that the first of equations (A) holds throughout the entire process of motion, we obtain

$$W = \int \epsilon X \, dx = \int_0^v \beta^3 v \, dv = \mu V^2 \left\{ \frac{1}{\sqrt{1 - \left(\frac{v}{V}\right)^2}} - 1 \right\}.$$

Thus, W becomes infinitely large when $v = V$. As is the case for our previous results, superluminal velocities are not possible.

By virtue of the argument presented above, this expression for kinetic energy must also be valid for ponderable masses.

Let us now enumerate the properties of the electron's motion resulting from the system of equations (A) that are accessible to experiment.

1. From the second equation of the system (A) it follows that an electric force Y and a magnetic force N have an equally strong deflective effect on an electron moving with velocity v if $Y = Nv/V$. Thus we see that, using our theory, it is possible to determine the velocity of the electron from the ratio of the magnetic deflection A_m to the electric deflection A_e for arbitrary velocities, by applying the law

$$\frac{A_m}{A_e} = \frac{v}{V}.$$

This relation can be tested experimentally since the velocity of the electron can also be measured directly, e.g., using rapidly oscillating electric and magnetic fields.

2. From the derivation of the kinetic energy of an electron it follows that the potential difference traversed by the electron and the velocity v that the electron acquires must be related by the equation

$$P = \int X \, dx = \frac{\mu}{\epsilon} V^2 \left\{ \frac{1}{\sqrt{1 - \left(\frac{v}{V}\right)^2}} - 1 \right\}.$$

3. We calculate the radius of curvature R of the path of the electron if a magnetic force N, acting perpendicularly to its velocity, is present (as the only deflecting force). From the second of equations (A) we obtain:

$$-\frac{d^2y}{dt^2} = \frac{v^2}{R} = \frac{\epsilon}{\mu} \frac{v}{V} N \cdot \sqrt{1 - \left(\frac{v}{V}\right)^2}$$

or

$$R = V^2 \frac{\mu}{\epsilon} \frac{\frac{v}{V}}{\sqrt{1 - \left(\frac{v}{V}\right)^2}} \cdot \frac{1}{N}.$$

These three relations are a complete expression of the laws by which, according to the theory presented here, the electron must move.

In conclusion, let me note that my friend and colleague M. Besso steadfastly stood by me in my work on the problem discussed here, and that I am indebted to him for several valuable suggestions.

(Annalen der Physik 17 [1905]: 891–921)

EDITORIAL NOTES

[1]In the 1913 reprint, a note was added after "be valid": "What is meant is, 'be valid in the first approximation.'" If Einstein did not write

the additional notes to this paper, the contents of some of the notes suggest that he was consulted.

[2] Einstein introduces the designations Ξ, H, Z for the coordinates for the x', y', x' axes of the moving system.

[3] In the 1913 reprint, the following note is appended to the end of this line: "The Lorentz transformation equations are more simply derivable directly from the condition that, as a consequence of these equations, the relation $\xi^2 + \eta^2 + \zeta^2 = V^2\tau^2 = 0$ shall have the other $x^2 + y^2 + z^2 - V^2t^2 = 0$ as a consequence."

[4] See previous note.

[5] This result later became known as "the clock paradox." In 1911, Langevin seems to have first introduced human travelers, leading to the alternate name, "the twin paradox."

[6] In the 1913 reprint, the following note is appended to the word "Unruhuhr": "In contrast to the 'pendulum clock,' which—from the physical standpoint—is a system, to which the earth belongs; this had to be excluded."

[7] This fraction should be $\frac{w_\eta}{w_\xi}$.

[8] The term "motional magnetic force" was introduced by Heaviside. Einstein later defined the "magnetomotive force" as the force acting on a unit of magnetic charge moving through an electric field. To the order of approximation used in the discussion of "electromotive force," the magnetomotive force is given by $-1/V\,[\mathbf{v}, \mathbf{E}]$, where $\mathbf{E} = (L, M, N)$, $\mathbf{v} = (v, 0, 0)$, and the bracket is a vector product.

[9] Corrected by Einstein in a reprint copy to "for $v = -V$, $v = -\infty$."

[10] In ibid., "the connecting line 'light source–observer'" was canceled and interlineated with "direction of motion."

[11] α should be φ.

[12] In a reprint copy, the denominator in the final term is corrected to "$1 - (v/V)^2$."

[13] In the 1913 reprint, the following note is appended to "call": "The definition of force given here is not advantageous as was first noted by M. Planck. It is instead appropriate to define force in such a way that the laws of momentum and energy conservation take the simplest form."

PAPER 4

✦

Does the Inertia of a Body
Depend on Its Energy Content?

THE RESULTS of an electrodynamic investigation recently published by me in this journal[1] lead to a very interesting conclusion, which will be derived here.

I based that investigation on the Maxwell-Hertz equations for empty space, together with Maxwell's expression for the electromagnetic energy of space, and also the following principle:

The laws according to which the states of physical systems change are independent of which one of the two coordinate systems (assumed to be in uniform parallel-translational motion relative to each other) is used to describe these changes (the principle of relativity).

Based on this foundation,[2] I derived the following result, among others (*loc. cit.*, sec. 8).

---

[1] A. Einstein, *Ann. d. Phys.* 17 (1905): 891. [See paper 3].

[2] The principle of the constancy of the velocity of light used there is of course contained in Maxwell's equations.

Let a system of plane light waves have the energy l relative to the coordinate system (x, y, z); let the ray direction (the wave-normal) make the angle φ with the x-axis of the system. If we introduce a new coordinate system (ξ, η, ζ), which is in uniform parallel translation with respect to the system (x, y, z), and the origin of which moves along the x-axis with velocity v, then this quantity of light—measured in the system (ξ, η, ζ)—has the energy

$$l^* = l \frac{1 - \frac{v}{V} \cos \varphi}{\sqrt{1 - \left(\frac{v}{V}\right)^2}},$$

where V denotes the velocity of light. We shall make use of this result in what follows.

Let there be a body at rest in the system (x, y, z) whose energy, relative to the system (x, y, z), is E_0. Let the energy of the body be H_0, relative to the system (ξ, η, ζ), moving with velocity v as above.

Let this body emit plane light waves of energy $L/2$ (measured relative to (x, y, z)) in a direction forming an angle φ with the x-axis, and at the same time an equal amount of light in the opposite direction. The body remains at rest with respect to system (x, y, z) during this process. This process must satisfy the principle of conservation of energy, and must be true (according to the principle of relativity) with respect to both coordinate systems. If E_1 and H_1 denote the energy of the body after emission of the light, measured relative to the system (x, y, z) and the (ξ, η, ζ), respectively, we obtain, using the relation indicated above,

$$E_0 = E_1 + \left[\frac{L}{2} + \frac{L}{2}\right],$$

$$H_0 = H_1 + \left[\frac{L}{2} \frac{1 - \frac{v}{V} \cos \varphi}{\sqrt{1 - \left(\frac{v}{V}\right)^2}} + \frac{L}{2} \frac{1 + \frac{v}{V} \cos \varphi}{\sqrt{1 - \left(\frac{v}{V}\right)^2}} \right]$$

$$= H_1 + \frac{L}{\sqrt{1 - \left(\frac{v}{V}\right)^2}}.$$

By subtraction, we obtain from these equations

$$(H_0 - E_0) - (H_1 - E_1) = L \left\{ \frac{1}{\sqrt{1 - \left(\frac{v}{V}\right)^2}} - 1 \right\}.$$

Both differences of the form $H - E$ occurring in this expression have simple physical meanings. H and E are the energy values of the same body, related to two coordinate systems in relative motion, the body being at rest in one of the systems (system (x, y, z)). Hence it is clear that the difference $H - E$ can differ from the body's kinetic energy K with respect to the other system (system (ξ, η, ζ)) only by an additive constant C, which depends on the choice of the arbitrary additive constants in the energies H and E. We can therefore set

$$H_0 - E_0 = K_0 + C$$
$$H_1 - E_1 = K_1 + C,$$

since C does not change during the emission of light. So we get

$$K_0 - K_1 = L \left\{ \frac{1}{\sqrt{1 - \left(\frac{v}{V}\right)^2}} - 1 \right\}.$$

The kinetic energy of the body with respect to (ξ, η, ζ) decreases as a result of emission of the light by an amount that is independent of the properties of the body. Furthermore, the difference $K_0 - K_1$ depends on the velocity in the same way as does the kinetic energy of an electron (*loc. cit.*, sec. 10).

Neglecting magnitudes of the fourth and higher order, we can get[1]

$$K_0 - K_1 = \frac{L}{V^2} \frac{v^2}{2}.$$

From this equation one immediately concludes:

If a body emits the energy L in the form of radiation, its mass decreases by L/V^2. Here it is obviously inessential that the energy taken from the body turns into radiant energy, so we are led to the more general conclusion:

The mass of a body is a measure of its energy content; if the energy changes by L, the mass changes in the same sense by $L/9 \cdot 10^{20}$ if the energy is measured in ergs and the mass in grams.

It is not excluded that it will prove possible to test this theory using bodies whose energy content is variable to a high degree (e.g., radium salts).

If the theory agrees with the facts, then radiation carries inertia between emitting and absorbing bodies.

(Annalen der Physik 18 [1905]: 639–641)

EDITORIAL NOTES

[1]Einstein used the Newtonian limit of the body's kinetic energy in order to evaluate change in its rest mass.

Part Four

✦

Einstein's Early Work on the
Quantum Hypothesis

Einstein in Bern, ca. 1905. (Lotte Jacobi
Archives, University of New Hampshire)

In describing his 1905 papers, Einstein characterized only paper 5 as revolutionary (see the Introduction, p. 5). It is still regarded as revolutionary for challenging the unlimited validity of Maxwell's theory of light and suggesting the existence of light quanta. Paper 5 shows that, at a sufficiently high frequency, the entropy of equilibrium thermal (or "black-body") radiation behaves as if the radiation consisted of a gas of independent "quanta of light energy" (or simply "light quanta"), each with energy proportional to the frequency of the corresponding wave. Einstein showed how to explain several otherwise puzzling phenomena by assuming that the interaction of light with matter consists of the emission or absorption of such light quanta.

Einstein became familiar with black-body radiation well before 1905. Mach's *Wärmelehre*, which Einstein read in 1897 or shortly thereafter, contains two chapters on thermal radiation, culminating in a discussion of Gustav Robert Kirchhoff's work. Kirchhoff showed that the energy emission spectrum of a perfectly black body (defined as one that absorbs all incident radiation) at a given temperature is a universal function of the temperature of the body and wavelength of the radiation. He inferred that equilibrium thermal radiation in a cavity with walls maintained at a certain temperature behaves like radiation emitted by a black body at the same temperature.

Heinrich Friedrich Weber, Einstein's physics professor at the Eidgenössische Technische Hochschule (ETH), was one of those who attempted to determine the universal blackbody radiation function. He made measurements of the energy spectrum and proposed an empirical formula for the distribution function. He showed that, as a consequence of his formula, $\lambda_m = \text{constant}/T$ (where λ_m is the wavelength with the maximum intensity of the distribution), thus anticipating Wilhelm Wien's formulation of the displacement law for black-body radiation. Weber described his work in a course at the ETH given during the winter semester of 1898–1899, which Einstein took.

Einstein soon started to think seriously about the problem of radiation. By the spring of 1901, he was closely following Planck's work on black-body radiation. Originally, Planck had hoped to explain the irreversibility of physical processes by studying electromagnetic radiation; ultimately, he came to recognize that this could not be done without introducing statistical elements into the argument. In a series of papers between 1897 and 1900, Planck utilized Maxwell's electrodynamics to develop a theory of thermal radiation in interaction with one or more identical, charged harmonic oscillators within a cavity. He was only able to account for the radiation's irreversible approach to thermal equilibrium by employing methods analogous to those that Ludwig Boltzmann had used in kinetic theory. Planck introduced the notion of "natural" (that is, maximally disordered) radiation, which he defined by analogy with Boltzmann's definition of molecular chaos. Using Maxwell's theory, Planck derived a relation between the average energy $\bar{E}$ of a charged oscillator of frequency ν in equilibrium with thermal radiation and the energy density per unit frequency interval ρ_ν of the radiation at the same frequency:

$$\bar{E}_\nu = \frac{c^3}{8\pi\nu^2}\,\rho_\nu, \tag{1}$$

where c is the speed of light.

Planck calculated the average energy of an oscillator by making assumptions about the entropy of the oscillators that enabled him to derive Wien's law for the energy density of the black-body spectrum, which originally seemed well supported by the experimental evidence. But by the turn of the century new observations showed systematic deviations from Wien's law for large values of λT.

Planck presented a new energy density distribution formula that agreed closely with observations over the entire spectrum:[1]

$$\rho_\nu = \frac{8\pi h\nu^3}{c^3} \frac{1}{e^{h\nu/kT} - 1}. \tag{2}$$

In this expression, now known as Planck's law or Planck's formula, $k = R/N$ is Boltzmann's constant, where R is the gas constant, N is Avogadro's (or Loschmidt's) number, and h is a new constant (later called Planck's constant). To derive this formula, Planck calculated the entropy of the oscillators, using what Einstein later called "the Boltzmann principle": $S = k \ln W$, where S is the entropy of a macroscopic state of the system, the probability of which is W. Following Boltzmann, Planck took the probability of a state to be proportional to the number of "complexions," or possible microconfigurations of the system corresponding to that state. He calculated this number by dividing the total energy of the state into a finite number of elements of equal magnitude, and counting the number of possible ways of distributing these energy elements among the individual oscillators. If the size of the energy elements is set equal to $h\nu$, where ν is the frequency of an oscillator, an expression for the entropy of an oscillator results that leads to eq. (2).

Although Einstein expressed private misgivings about Planck's approach in 1901, he did not mention Planck or black-body radiation in his papers until 1904. A study of the foundations of statistical physics, which he undertook between 1902 and 1904, provided Einstein with the tools he needed to analyze Planck's derivation and explore its consequences. At least three elements of Einstein's "general molecular theory of heat" were central to his subsequent work on the quantum hypothesis: (1) the introduction of the canonical ensemble; (2) the interpretation of probability as it occurs in Boltzmann's principle; and (3) the study of energy fluctuations in thermal equilibrium.

1. In an analysis of the canonical ensemble, Einstein proved that the equipartition theorem (see the Introduction, p. 16) holds for any system in thermal equilibrium. In paper 5 he showed that, when applied to an ensemble of charged harmonic oscillators in equilibrium with thermal radiation, the equipartition theorem leads, via eq. (1), to a black-body distribution law now known as the Rayleigh-Jeans law:

$$\rho_\nu = \frac{8\pi\nu^2}{c^3}\, kT. \qquad (3)$$

Despite its rigorous foundation in classical physics, eq. (3) only agrees with the observed energy distribution for small values of ν/T; indeed, as Einstein noted, it implies an infinite total radiant energy.

2. In 1906, Einstein posed a question that preoccupied him and others at the time: "How is it that Planck did not arrive at the same formula [eq. (3)], but at the expression . . . [eq. (2)]?" One answer lies in Planck's definition of W in Boltzmann's principle, which, as Einstein repeatedly noted, differs fundamentally from his own definition of probabilities as time averages. As noted above, Planck interpreted W as

proportional to the number of complexions of a state of the system. As Einstein pointed out in 1909, such a definition of W is equivalent to its definition as the average, over a long period of time, of the fraction of time that the system spends in this state only if all complexions corresponding to a given total energy are equally probable. However, if this is assumed to hold for an ensemble of oscillators in thermal equilibrium with radiation, the Rayleigh-Jeans law results. Hence, the validity of Planck's law implies that all complexions cannot be equally probable. Einstein showed that, if the energies available to a canonical ensemble of oscillators are arbitrarily restricted to multiples of the energy element $h\nu$, then all possible complexions are not equally probable, and Planck's law results.

3. A third element of Einstein's work on statistical physics that is central to his work on the quantum hypothesis is his method for calculating mean-square fluctuations in the state variables of a system in thermal equilibrium. He employed the canonical ensemble to calculate such fluctuations in the energy of mechanical systems, then boldly applied the result to a nonmechanical system—black-body radiation, deducing a relation that agrees with Wien's displacement law discussed above. This agreement suggests the applicability of statistical concepts to radiation, and may have suggested to Einstein the possibility that radiation could be treated as a system with a finite number of degrees of freedom, a possibility he raised at the outset of paper 5.

In connection with his work on Brownian motion in 1905–1906, Einstein developed additional methods for calculating fluctuations, methods he later applied to the analysis of black-body radiation. In particular, he developed a method based on an inversion of Boltzmann's principle that can be used even in the absence of a microscopic model of the system. If the entropy of a system is given as a function of its macroscopic state variables, then Boltzmann's principle, in the form $W = \exp(S/k)$, can be used to calculate the probability of a state, and hence of fluctuations of any state variable. In 1909, Einstein used this method to calculate the fluctuations in the energy of black-body radiation in a given region of space. The stochastic method, used in the same paper to calculate fluctuations in the pressure of radiation, is based on his work on Brownian motion. Pressure fluctuations maintain the Brownian motion of a small mirror moving through the radiation field, in the face of the retarding force exerted on the mirror by the average radiation pressure. The results of these fluctuation calculations are discussed below.

Einstein's work on relativity also contributed to the development of his views on the nature of light. By eliminating the concept of the ether and showing that a flux of radiant energy transfers inertial mass, the theory of relativity demonstrated that light no longer need be treated as a disturbance in a hypothetical medium, but could be regarded as composed of independent structures, to which mass must be attributed.

AMONG Einstein's papers on the quantum hypothesis, paper 5 is unique in arguing for the notion of light quanta without using either the formal apparatus of his statistical papers or Planck's law. As noted above, Einstein demonstrated that only the Rayleigh-Jeans law, the limiting form of Planck's formula for small values of ν/T, is consistent with classical statistical mechanics and Maxwell's electrodynamics. At the other extreme, at which Wien's distribution law holds,

Einstein argues that "theoretical principles we have been using . . . fail completely." As he explained later that year, this failure "seems to me to have its basis in a fundamental incompleteness of our physical concepts."[2]

Einstein opened the paper by pointing out the "profound formal difference" between current theories of matter, in which the energy of a body is represented as a sum over a finite number of degrees of freedom, and Maxwell's theory, in which the energy is a continuous spatial function of fields having an infinite number of degrees of freedom. He suggested that the inability of Maxwell's theory to give an adequate account of radiation might be remedied by a theory in which radiant energy is distributed discontinuously in space. Einstein formulated "the light quantum hypothesis" that "when a light ray spreads out from a point source, the energy is not distributed continuously over an ever-increasing volume but consists of a finite number of energy quanta that are localized at points in space, move without dividing, and can be absorbed or generated only as complete units."

Using Wien's law, Einstein showed that the expression for the volume dependence of the entropy of radiation at a given frequency is similar in form to that of the entropy of an ideal gas. He concluded that "monochromatic radiation of low density (within the range of validity of Wien's radiation formula) behaves thermodynamically as if it consisted of mutually independent energy quanta of magnitude $R\beta v/N$."

In addition to its contributions to theory, paper 5 also provides ingenious explanations of several observed phenomena. It examines three interactions of light with matter, treated "as if light consisted of such energy quanta": Stokes's rule for fluorescence; the ionization of gases by ultraviolet light; and the photoelectric effect. Einstein proposed what later became known as his photoelectric equation,

$$E_{\max} = (R/N)\beta\nu - P, \qquad (4)$$

where $E_{\max}$ is the maximum kinetic energy of the photoelectrons, $R\beta/N$ is equivalent to Planck's h, ν is the frequency of the incident radiation, and P is the work function of the metal emitting the electrons. Although his derivation of this equation was later considered to be a leading achievement of that paper—it is cited in his 1922 Nobel Prize award—for almost two decades this argument failed to persuade most physicists of the validity of the light quantum hypothesis. Philip Lenard's experimental studies, to which Einstein referred, only provide qualitative evidence for an increase of $E_{\max}$ with frequency. For almost a decade the quantitative relationship between electron energy and radiation frequency was in doubt. By about 1914, there was a substantial body of evidence tending to support eq. (4). Robert Millikan's 1916 studies clinched the case for almost all physicists. But even the confirmation of Einstein's photoelectric equation did not bring about widespread acceptance of the concept of light quanta. Alternative interpretations of the photoelectric effect still received general support for a number of years.

The earliest widely accepted empirical evidence for the quantum hypothesis came not from radiation phenomena, but from data on specific heats of solids. In 1907, Einstein applied the quantum hypothesis to the model of a solid as a lattice of atoms harmonically bound to their equilibrium positions.[3] When the oscillators are treated classically, the equipartition theorem leads to the DuLong-Petit rule, predicting a constant specific heat for solids at all temperatures. Treating each atom as a quantized three-dimensional harmonic oscillator, Einstein was able to explain the well-known anomalous decrease of the specific heats of certain solids with decreasing temperature, and to obtain a relation between the specific heat of a solid and its selective absorption of infrared radiation.

Einstein had long suspected the existence of such a connection. Perhaps inspired by Planck's work, in 1901 Einstein wondered whether the internal kinetic energy of solids and liquids could be conceived of as "the energy of electric resonators." If it could, then the "specific heat and absorption spectrum of bodies must then be related."[4] He tried to connect this model with deviations from the DuLong-Petit rule.

In 1907 Einstein avoided the implications of the equipartition theorem for the specific heats of solids as he had avoided them for radiation theory, by introducing energy quanta. From the average energy of a quantized oscillator, Einstein derived an expression for the specific heat of a monatomic solid as a function of $\beta T/\nu$. The expression approaches zero with the temperature, and approaches the DuLong-Petit value at high temperatures. Considering the simplified nature of the model, the expression is in fairly good agreement with Weber's data on diamond.

A connection could also be made with absorption results from Drude's optical dispersion theory. Drude showed that the infrared optical eigenfrequencies of a solid are due to vibrations of the lattice ions, while the electrons are responsible for the ultraviolet eigenfrequencies. At room temperature, the value of Einstein's expression for specific heat effectively vanishes at a frequency well within the infrared region for most solids, and increases to the DuLong-Petit value for even lower frequencies. Einstein concluded that only the lattice ions and atoms contribute to the specific heats of solids. Moreover, if a solid displays infrared absorption resonances, the temperature dependence of its specific heat can be determined from these resonant frequencies. In 1910 Nernst and his assistant Frederick A. Lindemann obtained general agreement between Einstein's predictions and observations of the variation with temperature of the specific heat of a number of solids. In 1911 Nernst reported the first confirmation of the quantum hypothesis outside the field of radiation, observing: "That the observations in their totality provide a brilliant confirmation of the quantum theory of Planck and Einstein is obvious."[5]

EDITORIAL NOTES

[1]Planck, *Annalen der Physik* 1 (1900): 719–737.

[2]Einstein, "Zur Theorie der Brownschen Bewegung," *Collected Papers*, vol. 2, doc. 32, pp. 334–345.

[3]"Planck's Theory of Radiation and the Theory of Specific Heats," *Annalen der Physik* 22 (1907): 180–190, reprinted in *Collected Papers*, vol. 2, doc. 38, pp. 379–389.

[4]Albert Einstein to Mileva Marić, 23 March 1901, *Collected Papers*, vol. 1, doc. 93.

[5]"Untersuchungen über die spezifische Wärme bei tiefen Temperaturen. III," *Königlich Preussische Akademie der Wissenschaften* (Berlin), *Sitzungsberichte* (1911), p. 310.

PAPER 5

✦

On a Heuristic Point of View
Concerning the Production and
Transformation of Light

A PROFOUND formal difference exists between the theoretical concepts that physicists have formed about gases and other ponderable bodies, and Maxwell's theory of electromagnetic processes in so-called empty space. While we consider the state of a body to be completely determined by the positions and velocities of an indeed very large yet finite number of atoms and electrons, we make use of continuous spatial functions to determine the electromagnetic state of a volume of space, so that a finite number of quantities cannot be considered as sufficient for the complete determination of the electromagnetic state of space. According to Maxwell's theory, energy is considered to be a continuous spatial function for all purely electromagnetic phenomena, hence also for light, whereas according to the present view of physicists, the energy of a ponderable body should be represented as a sum over the atoms and electrons. The energy of a ponderable body cannot be broken up into arbitrarily many, arbitrarily small parts, but according to Maxwell's theory (or, more generally, according to any wave theory) the energy of a light ray emitted from a point source continuously spreads out over an ever-increasing volume.

The wave theory of light, which operates with continuous spatial functions, has proved itself superbly in describing purely optical phenomena and will probably never be replaced by another theory. One should keep in mind, however, that optical observations refer to time averages rather than instantaneous values; and it is quite conceivable, despite the complete confirmation of the theory of diffraction, reflection, refraction, dispersion, etc., by experiment, that the theory of light, operating with continuous spatial functions, leads to contradictions when applied to the phenomena of emission and transformation of light.

Indeed, it seems to me that the observations of "blackbody radiation," photoluminescence, production of cathode rays by ultraviolet light, and other related phenomena associated with the emission or transformation of light appear more readily understood if one assumes that the energy of light is discontinuously distributed in space. According to the assumption considered here, in the propagation of a light ray emitted from a point source, the energy is not distributed continuously over ever-increasing volumes of space, but consists of a finite number of energy quanta localized at points of space that move without dividing, and can be absorbed or generated only as complete units.

In this paper I wish to present the train of thought and cite the facts that led me onto this path, in the hope that the approach to be presented will prove of use to some researchers in their investigations.

1. ON A DIFFICULTY CONCERNING
THE THEORY OF
"BLACK-BODY RADIATION"

We shall begin by considering the following case from the perspective of Maxwell's theory and electron theory. Let a space enclosed by completely reflecting walls contain a number of freely moving gas molecules and electrons that exert conservative forces on each other when they come very close, i.e., can collide with each other like molecules according to the kinetic theory of gases.[1] Suppose, further, that a number of electrons are bound to widely separated points in space by forces directed toward these points and proportional to their distances from them. The bound electrons also enter into conservative interactions with the free molecules and electrons when the latter come very close. We call the bound electrons "resonators"; they emit and absorb electromagnetic waves of definite periods.

According to the present view concerning the origin of light, the radiation in the volume we are considering, as is found for the case of dynamic equilibrium based on Maxwell's theory, must be identical with "black-body radiation"—at least if one assumes that resonators of all relevant frequencies are present.

For the time being, we will disregard the radiation emitted and absorbed by the resonators and investigate the condition of dynamic equilibrium corresponding to the interac-

---

[1] This assumption is equivalent to the supposition that the mean kinetic energies of gas molecules and electrons are equal to each other at thermal equilibrium. As is well known, Mr. Drude used the latter assumption to derive a theoretical expression for the ratio of thermal and electrical conductivities of metals.

tion (collisions between) molecules and electrons. For such an equilibrium, the kinetic theory of gases asserts that the mean kinetic energy of a resonator electron must be equal to the mean translational kinetic energy of a gas molecule. If we decompose the motion of a resonator electron into three mutually perpendicular oscillatory motions, we find for the mean value $\bar{E}$ of the energy of one such linear oscillatory motion

$$\bar{E} = \frac{R}{N}T,$$

where R denotes the universal gas constant, N the number of "real molecules"[1] in one gram-equivalent, and T the absolute temperature. Because of the equality of the time averages of the resonator's kinetic and potential energies, the energy $\bar{E}$ is two-thirds the kinetic energy of a free monatomic gas molecule. If due to some cause—in our case, radiation processes—the energy of a resonator were to have a time average greater or less than $\bar{E}$, then collisions with the free electrons and molecules would lead to an energy loss to or a gain from the gas, different on the average from zero. Thus, in the case we are considering, dynamic equilibrium is possible only if the average energy of each resonator is $\bar{E}$.

We will now apply a similar argument to the interaction between the resonators and the radiation present in space. Mr. Planck has derived the condition for dynamic equilibrium in this case[2] on the assumption that the radiation can be treated as the most completely conceivable disordered

[2] M. Planck, *Ann. d. Phys.* 1 (1900): 99.

process.[3] He found:

$$\bar{E}_\nu = \frac{L^3}{8\pi\nu^2}\rho_\nu.$$

$\bar{E}_\nu$ is here the mean energy of a resonator with eigenfrequency ν (per unit frequency interval), L is the velocity of light, ν the frequency, and $\rho_\nu\, d\nu$ the energy per unit volume of that part of the radiation whose frequency lies between ν and $\nu + d\nu$.

[3] This assumption can be formulated as follows. We expand the Z-component of the electrical force (Z) at an arbitrary point of the space being considered during the time interval between $t = 0$ and $t = T$ (where T denotes a time period very large relative to all relevant oscillation periods) in a Fourier series,

$$Z = \sum_{\nu=1}^{\nu=\infty} A_\nu \sin\left(2\pi\nu\frac{t}{T} + \alpha_\nu\right),$$

where $A_\nu \geq 0$ and $0 \leq \alpha_\nu \leq 2\pi$. If one imagines that at the same point of space such an expansion is made arbitrarily often at randomly chosen starting times, then one will obtain different sets of values for the quantities A_ν and α_ν. For the frequency of occurrence of the various combinations of values of the quantities A_ν and α_ν, there will then exist (statistical) probabilities dW of the form

$$dW = f(A_1 A_2 \ldots \alpha_1 \alpha_2 \ldots)\, dA_1 dA_2 \ldots d\alpha_1 d\alpha_2 \ldots$$

The radiation is then most conceivably disordered if

$$f(A_1, A_2 \ldots \alpha_1, \alpha_2 \ldots) = F_1(A_1)F_2(A_2)\ldots f_1(\alpha_1)\cdot f_2(\alpha_2)\ldots,$$

i.e., when the probability of a specific value of one of the quantities A or $x^{[2]}$ is independent of the values taken by the other quantities A and α respectively. Hence, the more closely fulfilled the condition is that individual pairs of quantities A_ν and α_ν depend on the emission and absorption processes of particular groups of resonators, the more closely can radiation in our case be considered to be "most conceivably disordered."

If, on the whole, the radiation energy of frequency ν does not continually decrease or increase, the following relations must hold:

$$\frac{R}{N}T = \bar{E} = \bar{E}_\nu = \frac{L^3}{8\pi\nu^2}\rho_\nu,$$

$$\rho_\nu = \frac{R}{N}\frac{8\pi\nu^2}{L^3}T. \text{[3]}$$

These relations, found as the conditions of dynamic equilibrium, not only fail to agree with experiment but also imply that in our model a definite distribution of energy between ether and matter is out of the question. Indeed, the wider the range of the resonator frequencies chosen, the greater the total radiation energy of the volume, and in the limit we obtain

$$\int_0^\infty \rho_\nu \, d\nu = \frac{R}{N}\frac{8\pi}{L^3}T\int_0^\infty \nu^2 \, d\nu = \infty.$$

2. On Planck's Determination of the Elementary Quanta[4]

We now wish to show that Mr. Planck's determination of the elementary quanta is to a certain extent independent of his theory of "black-body radiation."

Planck's formula[4] for ρ_ν, which is satisfied by all experiments up to now, reads

$$\rho_\nu = \frac{\alpha\nu^3}{e^{\frac{\beta\nu}{T}} - 1},$$

[4] M. Planck, *Ann. d. Phys.* 4 (1901): 561.

where

$$\alpha = 6.10 \times 10^{-56}$$
$$\beta = 4.866 \times 10^{-11}.$$

For large values of T/ν, i.e., for high wavelengths and radiation densities, this equation takes the following limiting form:

$$\rho_\nu = \frac{\alpha}{\beta}\nu^2 T.$$

One can see that this formula agrees with that derived in section 1 from Maxwell's theory and electron theory. By equating the coefficients of the two formulas, we obtain

$$\frac{R}{N}\frac{8\pi}{L^3} = \frac{\alpha}{\beta}$$

or

$$N = \frac{\beta}{\alpha}\frac{8\pi R}{L^3} = 6.17 \times 10^{23},$$

i.e., one atom of hydrogen weighs $1/N$ gram $= 1.62 \times 10^{-24}$ g. This is exactly the value found by Mr. Planck, which in turn is in satisfactory agreement with the values found by other methods.

We therefore arrive at the following conclusion: the higher the energy density and wavelength of radiation, the more reasonable the theoretical foundations we have been using prove to be; however, they fail completely in the case of low wavelengths and low radiation densities.

In the following, we shall consider "black-body radiation" together with the experimental facts, without establishing any model for the emission and propagation of radiation.

3. On the Entropy of Radiation

The following treatment is contained in a well-known study by Mr. Wien and is presented here only for the sake of completeness.

Consider radiation occupying a volume v. We assume that the observable properties of this radiation are completely determined when the radiation density $\rho(\nu)$ is given for all frequencies.[5] Since radiations of different frequencies can be regarded as separable from each other without performing any work or transferring any heat, the entropy of the radiation can be represented by

$$S = v \int_0^\infty \varphi(\rho, \nu) \, d\nu,$$

where φ is a function of the variables ρ and ν. One can reduce φ to a function of a single variable by asserting that adiabatic compression of radiation between reflecting walls does not change its entropy. However, we shall not pursue this, but will immediately investigate how the function φ can be obtained from the black-body radiation law.

In the case of "black-body radiation," ρ is such a function of ν that the entropy is a maximum for a given energy, i.e.,

$$\delta \int_0^\infty \varphi(\rho, \nu) \, d\nu = 0$$

if

$$\delta \int_0^\infty \rho \, d\nu = 0.$$

---

[5] This assumption is arbitrary. We shall naturally keep this simplest assumption as long as experiment does not force us to abandon it.

From this it follows that for every choice of $\delta\rho$ as a function of ν,

$$\int_0^\infty \left(\frac{\partial\varphi}{\partial\rho} - \lambda \right) \delta\rho \, d\nu = 0,$$

where λ is independent of ν. Thus, for black-body radiation, $\partial\varphi/\partial\rho$ is independent of ν.

The following equation applies when the temperature of black-body radiation of volume $v = 1$ increases by dT:

$$dS = \int_{\nu=0}^{\nu=\infty} \frac{\partial\varphi}{\partial\rho} \, d\rho \, d\nu,$$

or, since $\partial\varphi/\partial\rho$ is independent of ν,

$$dS = \frac{\partial\varphi}{\partial\rho} \, dE.$$

Since dE is equal to the heat added and the process is reversible, we also have

$$dS = \frac{1}{T} \, dE.$$

Comparison yields

$$\frac{\partial\varphi}{\partial\rho} = \frac{1}{T}.$$

This is the law of black-body radiation. Thus, one can derive the law of black-body radiation from the function φ, and, inversely, the function φ can be determined by integration, remembering that φ vanishes for $\rho = 0$.

4. LIMITING LAW FOR THE ENTROPY OF MONOCHROMATIC RADIATION AT LOW RADIATION DENSITY

The existing observations of "black-body radiation" show that the law

$$\rho = \alpha \nu^3 e^{-\beta \frac{\nu}{T}}$$

originally postulated by Mr. W. Wien for "black-body radiation" is not strictly valid. However, it has been fully confirmed by experiment for large values of ν/T. We shall base our calculations on this formula, bearing in mind, however, that our results are valid only within certain limits.

First of all, this formula yields

$$\frac{1}{T} = -\frac{1}{\beta \nu} \ln \frac{\rho}{\alpha \nu^3},$$

and next, using the relation obtained in the preceding section,

$$\varphi(\rho, \nu) = -\frac{\rho}{\beta \nu} \left\{ \ln \frac{\rho}{\alpha \nu^3} - 1 \right\}.$$

Suppose now that we have radiation of energy E, with frequency between ν and $\nu + d\nu$, occupying a volume v. The entropy of this radiation is

$$S = v\varphi(\rho, \nu)\, d\nu = -\frac{E}{\beta \nu} \left\{ \ln \frac{E}{v\alpha \nu^3 d\nu} - 1 \right\}. \quad [5]$$

If we restrict ourselves to investigating the dependence of the entropy on the volume occupied by the radiation, and denote the entropy of radiation by S_0 at volume v_0, we obtain

$$S - S_0 = \frac{E}{\beta \nu} \ln \left[\frac{v}{v_0} \right].$$

This equation shows that the entropy of monochromatic radiation of sufficiently low density varies with the volume according to the same law as the entropy of an ideal gas or a dilute solution. In the following, we shall interpret this equation on the basis of the principle introduced into physics by Mr. Boltzmann, according to which the entropy of a system is a function of the probability of its state.

5. Molecular-Theoretical Investigation of the Dependence on Volume of the Entropy of Gases and Dilute Solutions

In calculating entropy by molecular-theoretical methods, the word "probability" is often used in a sense differing from the way the word is defined in probability theory. In particular, "cases of equal probability" are often hypothetically stipulated when the theoretical models employed are definite enough to permit a deduction rather than a stipulation. I will show in a separate paper that, in dealing with thermal processes, it suffices completely to use the so-called statistical probability, and hope thereby to eliminate a logical difficulty that still obstructs the application of Boltzmann's principle. Here, however, I shall give only its general formulation and its application to very special cases.

If it is meaningful to talk about the probability of a state of a system, and if, furthermore, each increase of entropy can be conceived as a transition to a state of higher probability, then the entropy S_1 of a system is a function of the probability W_1 of its instantaneous state. Therefore, if we have two systems S_1 and S_2 that do not interact with each other, we can set:

$$S_1 = \varphi_1(W_1),$$
$$S_2 = \varphi_2(W_2).$$

If these two systems are viewed as a single system of entropy S and probability W, then

$$S = S_1 + S_2 = \varphi(W)$$

and

$$W = W_1 \cdot W_2.$$

The last equation tells us that the states of the two systems are events that are independent of each other.

From these equations it follows that

$$\varphi(W_1 \cdot W_2) = \varphi_1(W_1) + \varphi_2(W_2),$$

and, finally,

$$\varphi_1(W_1) = C \ln(W_1) + \text{const.}$$
$$\varphi_2(W_2) = C \ln(W_2) + \text{const.}$$
$$\varphi(W) = C \ln(W) + \text{const.}$$

The quantity C is therefore a universal constant; it follows from the kinetic theory of gases that its value is R/N, where the meaning of the constants R and N is the same as above. If S_0 denotes the entropy of a certain initial state and W is the relative probability of a state of entropy S, we obtain in general:

$$S - S_0 = \frac{R}{N} \ln W.$$

We now treat the following special case. Let a volume v_0 contain a number (n) of moving points (e.g., molecules) to which our discussion will apply. The volume may also contain any arbitrary number of other moving points of any kind. No assumption is made about the law governing the motion of the points under discussion in the volume except that, as concerns this motion, no part of the space (and no direction within it) is preferred over the others. Further, let the number of (aforementioned) moving points under discussion be small enough that we can disregard interactions between them.

This system, which, for example, can be an ideal gas or a dilute solution, possesses a certain entropy S_0. Let us imagine that all n moving points are assembled in a part of the volume v_0 of size v without any other changes in the system. It is obvious that this state has a different value of entropy (S), and we now wish to determine the difference in entropy with the help of Boltzmann's principle.

We ask: How great is the probability of the last-mentioned state relative to the original one? Or: How great is the probability that at a randomly chosen instant of time, all n independently moving points in a given volume v_0 will be found (by chance) in the volume v?

Obviously, this probability, which is a "statistical probability," has the value

$$W = \left(\frac{v}{v_0}\right)^n;$$

from this, by applying Boltzmann's principle, one obtains

$$S - S_0 = R\left(\frac{n}{N}\right)\ln\left(\frac{v}{v_0}\right).$$

It is noteworthy that, in the derivation of this equation, from which the Boyle–Gay-Lussac law and the analogous law of osmotic pressure can easily be derived thermo-dynamically,[6] no assumptions need be made about the law governing the motion of the molecules.

6. INTERPRETATION ACCORDING TO BOLTZMANN'S PRINCIPLE OF THE EXPRESSION FOR THE DEPENDENCE OF THE ENTROPY OF MONOCHROMATIC RADIATION ON VOLUME

In section 4 we found the following expression for the dependence of the entropy of monochromatic radiation on volume:

$$S - S_0 = \frac{E}{\beta \nu} \ln\left(\frac{v}{v_0}\right).$$

If we write this formula in the form

$$S - S_0 = \frac{R}{N} \ln\left[\left(\frac{v}{v_0}\right)^{\frac{N}{R}\frac{E}{\beta \nu}}\right]$$

and compare it with the general formula expressing the Boltzmann principle,

$$S - S_0 = \frac{R}{N} \ln W,$$

we arrive at the following conclusion: If monochromatic radiation of frequency ν and energy E is enclosed (by reflecting

---

[6] If E is the energy of the system, we get

$$-d(E - TS) = pdv = T\, dS = R\frac{n}{N}\frac{dv}{v};^{[6]}$$

thus

$$pv = R\,\frac{n}{N}T.$$

walls) in the volume v_0, the probability that at a randomly chosen instant the total radiation energy will be found in the portion v of the volume v_0 is

$$W = \left(\frac{v}{v_0}\right)^{\frac{N}{R}\frac{E}{\beta\nu}}.$$

From this we further conclude that monochromatic radiation of low density (within the range of validity of Wien's radiation formula) behaves thermodynamically as if it consisted of mutually independent energy quanta of magnitude $R\beta\nu/N$.[7]

We also want to compare the mean value of the energy quanta of black-body radiation with the mean kinetic energy of the center-of-mass motion of a molecule at the same temperature. The latter is $\frac{3}{2}(R/N)T$, while the mean value of the energy quantum obtained on the basis of Wien's formula is

$$\frac{\displaystyle\int_0^\infty \alpha\nu^3 e^{-\frac{\beta\nu}{T}}\, d\nu}{\displaystyle\int_0^\infty \frac{N}{R\beta\nu}\alpha\nu^3 e^{-\frac{\beta\nu}{T}}\, d\nu} = 3\frac{R}{N}T.$$

If monochromatic radiation (of sufficiently low density) behaves, as concerns the dependence of its entropy on volume, as though the radiation were a discontinuous medium consisting of energy quanta of magnitude $R\beta\nu/N$, then it seems reasonable to investigate whether the laws governing the emission and transformation of light are also constructed as if light consisted of such energy quanta. We will consider this question in the following sections.

7. On Stokes's Rule

Let monochromatic light be transformed by photoluminescence into light of a different frequency, and, in accordance
with the result just obtained, let us assume that both the
incident and emitted light consist of energy quanta of magnitude $(R/N)\beta\nu$, where ν denotes the relevant frequency. The
transformation process can then be interpreted as follows.
Each incident energy quantum of frequency ν_1 is absorbed
and—at least at a sufficiently low distribution density of the
incident energy quanta—generates by itself a light quantum
of frequency ν_2; it is possible that the absorption of the
incident light quantum may simultaneously give rise to
the emission of light quanta of frequencies ν_3, ν_4, etc., as
well as to energy of some other kind (e.g., heat). It makes no
difference by means of what intermediary processes this
end result occurs. If the photoluminescent substance is not
regarded as a permanent energy source, then, according to
the principle of conservation of energy, the energy of an
emitted energy quantum cannot be greater than that of the
light quantum that produced it; hence it follows that

$$\frac{R}{N}\beta\nu_2 \leqq \frac{R}{N}\beta\nu_1,$$

or

$$\nu_2 \leqq \nu_1.$$

This is the well-known Stokes's rule.

It is to be emphasized that, according to our conception,
at low illumination the amount of light emitted must be
proportional to the intensity of the incident light, because
each incident energy quantum will induce an elementary
process of the kind indicated above, independently of other

incident energy quanta. In particular, there will be no lower limit for the intensity of the incident light, below which the light would be unable to excite the fluorescent effect.

According to the conception of the phenomenon presented here, deviations from Stokes's rule are conceivable in the following cases:

1. If the number of energy quanta per unit volume simultaneously being transformed is so large that an energy quantum of emitted light can obtain its energy from several incident energy quanta.

2. When the incident (or emitted) light does not have the same energy distribution as black-body radiation within the range of validity of Wien's law; if, for example, the incident light is produced by a body of such high temperature that Wien's law is no longer valid for the relevant wavelengths.

The latter possibility deserves special attention. According to the conception developed above, it is indeed not impossible that even at very low densities a "non-Wien radiation" will behave differently as concerns its energy from black-body radiation within the range of validity of Wien's law.

8. On the Generation of Cathode Rays by Illumination of Solid Bodies

The usual view that the energy of light is continuously distributed over the space through which it travels faces especially great difficulties when one attempts to explain photoelectric phenomena, which are expounded in a pioneering work by Mr. Lenard.[7]

[7] P. Lenard, *Ann. d. Phys.* 8 (1902): 169, 170.

According to the view that the incident light consists of energy quanta of energy $(R/N)\beta\nu$, the production of cathode rays by light can be conceived in the following way. The body's surface layer is penetrated by energy quanta whose energy is converted at least partially into kinetic energy of the electrons. The simplest conception is that a light quantum transfers its entire energy to a single electron; we will assume that this can occur. However, we will not exclude the possibility that electrons absorb only a part of the energy of the light quanta.

An electron in the interior of the body having kinetic energy will have lost part of its kinetic energy by the time it reaches the surface. In addition, we will assume that, in leaving the body, each electron must perform some work, P (characteristic for the body). The electrons leaving the body with the greatest perpendicular velocity will be those located right on the surface and ejected normally to it. The kinetic energy of such electrons is

$$\frac{R}{N}\beta\nu - P.$$

If the body is charged to a positive potential Π and surrounded by conductors at zero potential, and if Π is just sufficient to prevent a loss of electrical charge by the body, it follows that

$$\Pi\epsilon = \frac{R}{N}\beta\nu - P,$$

where ϵ denotes the charge of the electron; or

$$\Pi E = R\beta\nu - P',$$

where E is the charge of one gram-equivalent of a monovalent ion and P' is the potential of this quantity of negative charge relative to the body.[8]

If one sets $E = 9.6 \times 10^3$, then $\Pi \cdot 10^{-8}$ is the potential in volts that the body acquires when irradiated in vacuum.

To see whether the derived relation agrees in order of magnitude with experiment, set $P' = 0$, $\nu = 1.03 \times 10^{15}$ (which corresponds to the ultraviolet limit of the solar spectrum) and $\beta = 4.866 \times 10^{-11}$. We obtain $\Pi \cdot 10^7 = 4.3$ volts, a result that agrees in order of magnitude with the results of Mr. Lenard.[9]

If the formula derived is correct, then Π, when plotted in Cartesian coordinates as a function of the frequency of the incident light, must give a straight line whose slope is independent of the nature of the substance under study.

As far as I can tell, this conception of the photoelectric effect does contradict its properties as observed by Mr. Lenard. If each energy quantum of the incident light transmits its energy to electrons, independently of all others, then the velocity distribution of the electrons, i.e., the nature of cathode rays produced, will be independent of the intensity of the incident light; on the other hand, under otherwise identical circumstances, the number of electrons leaving the body will be proportional to the intensity of the incident light.[10]

---

[8] If one assumes that the individual electron can only be detached by light from a neutral molecule by the expenditure of some work, one does not have to change anything in the relation just derived; one need only consider P' as the sum of two terms.

[9] P. Lenard, *Ann. d. Phys.* 8 (1902): 165, 184, table I, fig. 2.

[10] P. Lenard, *loc. cit.*, pp. 150, 166–168.

Remarks similar to those regarding possible deviations from Stokes's rule apply to the possible limits of validity of the laws set forth above.

In the foregoing it has been assumed that at least some of the quanta of the incident light transmit their energies completely to one electron each. If this likely assumption is not made, one obtains the following equation in place of the last one:

$$\Pi E + P' \leqq R\beta\nu.$$

For cathode luminescence, which is the inverse of the process discussed above, one obtains by analogous considerations:

$$\Pi E + P' \geqq R\beta\nu.$$

For the substances investigated by Mr. Lenard, PE[8] is always considerably greater than $R\beta\nu$ because the potential difference, which the cathode rays must traverse in order to produce visible light, amounts in some cases to hundreds, in others to thousands of volts.[11] We must therefore assume that the kinetic energy of one electron goes into the production of many energy quanta of light.

9. On the Ionization of Gases by Ultraviolet Light

We have to assume that, in ionization of a gas by ultraviolet light, one energy quantum of light serves to ionize one gas molecule. From this it follows that the ionization energy (i.e., the work theoretically required for its ionization) of a

[11] P. Lenard, *Ann. d. Phys.* 12 (1903): 469

molecule cannot be greater than the energy of an absorbed light quantum capable of producing this effect. If J denotes the (theoretical) ionization energy per gram-equivalent, it follows that

$$R\beta\nu \geq J.$$

According to Lenard's measurements, the largest effective wavelength for air is about 1.9×10^{-5}cm, hence

$$R\beta\nu = 6.4 \times 10^{12} \text{ erg} \geq J.$$

An upper limit for the ionization energy can also be obtained from the ionization potentials of rarefied gases. According to J. Stark[12] the smallest measured ionization potential (for platinum anodes) for air is about 10 volts.[13] Thus, one obtains 9.6×10^{12} as the upper limit for J, which is almost equal to the value found above. There is still another consequence, verification of which by experiment seems to me of great importance. If each absorbed energy quantum of light ionizes one molecule, then the following relation must hold between the quantity of light absorbed L and the number j of gram-molecules of ionized gas:

$$j = \frac{L}{R\beta\nu}.$$

If our conception is correct, this relation must be valid for all gases that (at the relevant frequency) show no appreciable absorption unaccompanied by ionization.

(*Annalen der Physik* 17 [1905]: 132–148)

[12] J. Stark, *Die Elektrizität in Gasen*, p. 57 (Leipzig, 1902).
[13] In the interior of the gases, however, the ionization potential of negative ions is five times greater.

EDITORIAL NOTES

[1]"Real molecules" are presumably those that are not dissociated.

[2]x should be α.

[3]Expressions equivalent to Einstein's equation were obtained in 1905 by Rayleigh and Jeans without the use of material resonators.

[4]Here "elementary quanta" refers to fundamental atomic constants. In 1901, Planck determined the mass of the hydrogen atom, Loschmidt's number (N), Boltzmann's constant, and the elementary electric charge.

[5]S refers to radiation with frequencies between ν and $\nu + d\nu$, and $E = \rho \nu \, d\nu$.

[6]The last term should be multiplied by T.

[7]$R\beta/N$ corresponds to Planck's "h."

[8]PE should be ΠE.